Project Management Demystified

Third Edition

Also available from Taylor & Francis

Programme Management Demystified   Pb: ISBN 978–0–419–21350–5

Construction Project Management   Hb: ISBN 978–0–415–35905–4
Peter Fewings   Pb: ISBN 978–0–415–35906–1

Risk Management in Projects   Hb: ISBN 978–0–415–26055–8
Martin Loosemore, John Raftery,   Pb: ISBN 978–0–415–26056–5
Charles Reilly, David Higgon

Practical Construction Management   Pb: ISBN 978–0–415–36257–3
R.H.B. Ranns & E.J.M. Ranns

Project Planning and Control   Hb: ISBN 978–0–415–34722–6
David G. Carmichael

Occupational Health & Safety in   Hb: ISBN 978–0–419–26210–7
Construction Project Management
Helen Lingard & Steve Rowlinson

Project Management Demystified

Third edition

GEOFF REISS

Taylor & Francis
Taylor & Francis Group

LONDON AND NEW YORK

First published 1992 by E & FN Spon
Reprinted 1992, 1993 (twice), 1994

Second edition 1995
Reprinted 1996, 1998, 1999

Reprinted by Spon Press 2000, 2001

Third edition published 2007 by Taylor & Francis
2 Park Square, Milton Park, Abingdon, Oxon OX14 4RN

Simultaneously published in the USA and Canada
by Taylor & Francis
270 Madison Ave, New York, NY 10016, USA

Taylor & Francis is an imprint of the Taylor & Francis Group, an informa business

© 1992, 1995, 2007 Geoff Reiss

Typeset in Palatino by
RefineCatch Limited, Bungay, Suffolk
Printed and bound in Great Britain by
MPG Books Ltd, Bodmin

British Library Cataloguing in Publication Data
A catalogue record for this book is available from the British Library

Library of Congress Cataloging in Publication Data
Reiss, Geoff, 1945–
Project management demystified / Geoff Reiss. — 3rd ed.
    p. cm.
Includes bibliographical references and index.
ISBN 978–0–415–42163–8 (pbk. : alk. paper) 1. Project management. I. Title.
HD69.P75R455 2007
658.4'04 – dc22
2007002562

ISBN10: 0–415–42163–2 (pbk)
ISBN10: 0–203–94572–7 (ebk)

ISBN-13: 978–0–415–42163–8 (pbk)
ISBN-13: 978–0–203–94572–8 (ebk)

Dedication

Without the help of my family,
Liz, Paul, Caroline, James, Caroline and Layla,
this book would have been finished many
years ago.

Contents

Acknowledgements

Many people have helped make this book what it is, whatever that is. Particularly, I would like to thank Paul Barton, The Construction Management Centre; Adrian Dooley, The Projects Group PLC; Robert Fearnley, Leeds; Dave Gove, Research Machines, Oxford; Ken Lane, Partnership PR, Wokingham; Ken and Sally Tomlinson, Hitchin; and Andy Wilkin, Pangbourne.

Preface

When my old grandfather came back from a long night watching over the local sardine factory he would often say 'You can never do well as a nightwatchman, my boy'. It took many years of hearing this seven times a week before I realized what he meant. If he did his nightwatchman job perfectly by staying awake all night, if no one broke in and there were no fires the result was that no one took any notice of him at all. One little problem, a minor break in, and he was to blame. He could easily do badly, he could never do well. He only got noticed if something went wrong.

Project management is a lot like that. The perfect project manager uses all the right techniques and tools to avoid problems so that nothing surprising happens and the project comes out well. No one, of course, takes any notice of the manager who achieved all that. Every problem, every delay, every overspend and everyone points the fickle finger of blame at the project manager.

For this reason *Project Management Demystified* aims to achieve two things. Firstly, you will get a peek into the project management toolbox – you will learn about critical path analysis, work breakdown structures, team building, cost control and a whole host of other simple, common sense techniques that help people bring projects in on time and to budget.

Successful projects do not always mean successful project managers, so the second aim of this book is to make you aware of opportunities to blow your own trumpet. It tells you how to make sure people recognize what a great project management person you really are!

You might be a manager, an engineer, a scientist or a student who thinks that project management has something to offer. You might have just been promoted to project manager when your boss dropped a project management software package in your lap. If you are in any of these categories read on, run good projects and have a successful career.

By the way, my grandfather was a tailor.

Geoff Reiss

Setting the stage

In this chapter the author explains a very personal view of project management and some of the popular misconceptions with which it is surrounded. In this chapter the Invisible Project Manager appears, and a plea is entered for equality among project managers.

A good project does not always make the project team look good. There are superbly run projects which came out bang on time and bang on budget but at the end of which the team got fired. There are terrible jobs where the thing finished years late, costing twice as much as the worst estimate, and yet the project manager was promoted out of sight. There are projects on which companies have lost lots of money but everyone thought that the company got off lightly. There are projects which were hugely profitable but which everyone thought were a complete mess. There are projects where devoted and capable people fight against enormous odds and achieve completely the wrong objective. There are projects where a team of clever people watch with their mouths wide open while some senior bureaucrat in a pin stripe suit gets all the credit.

Success in a project is not proportional to success in project management. All those heavy tomes on project management totally ignore this simple fact. I won't.

This book will cover the techniques that you should know before embarking on a project or for that matter on a career in project management. It will open up for you the project manager's toolkit. All these tools have to be handled by some intelligent life form with care and thought. The tools are mostly about looking forward and thinking ahead. Some are about measuring how things are going so that you can do more planning, more thinking ahead.

Within the humble pages of the following chapters are sections on critical path analysis, earned value analysis, work breakdown structures, and cash flow curves. Those of you who do read this book will read about PERT (Program Evaluation and Review Technique) and loads of other useful techniques and tools.

Certainly, you'll get a fighting chance of running a tidy project. This book will never forget that there are twin objectives in project management – the success of the project and of you; your future, your career, and your salary cheque.

This is a footnote. There will be more.

There are many twists and turns on the road to success in project management and some are very surprising indeed. One strategy you might adopt is to learn as much as you can about the topic and manage the project brilliantly. You might get professional help. You might meet all sorts of problems but, through quick thinking, overcome them. Such strategies can be turned to your own advantage and all can work against you.

If you get professional help at least you have someone else to blame – 'We got the best experts in and if they couldn't sort it out, what could we have done?' You might fail dismally, but this may not come to light until some time after you have left the project, so that someone else gets the blame. One great danger lies in doing too well.

Project management centres on avoiding problems

A great deal of project management involves avoiding problems. As you will read, project management is about tackling new ground, taking a group of people and trying to achieve some very clear objective quickly and efficiently.

Simply because the challenge is going to be new, there are going to be new problems. Such new problems are likely to demand exactly the things you are most short of: time and money. So, the more problems that you can foresee and solve before they start to eat into your precious reserves of time and money, the better.

Factory managers do not normally project manage. They practise running their factories all day long. They manage their factories and if there is a problem they are expected to overcome it by spending some time and money. We shall try to get a clear view of project management later on, but for the moment please accept that projects tend to be unpractised and unrehearsed.

For some reason, project management people are expected to avoid problems, despite the usually unknown nature of their work.

Problems remind me of ships coming into view over the horizon. While they are small and distant, there is time to do something about them. Drake could finish his game of bowls and prepare to defend England.

But if you do not spot those ships/problems, they grow and grow, becoming ever more threatening until they are finally impossible to avoid. When they bump right into your nose you cannot help noticing them, but by then they loom large. There is no time to solve them; the problems have become big and inescapable.

While you are busy doing your best to overcome a large, imminent problem, guess what is happening? Because you are not looking, lots of other nasty problems are taking the chance to creep closer. The more big problems you are dealing with, the less time you have to watch out for little ones getting bigger. You could try to deal with these problems while they are small and far away. By spending much time worrying about the future, you might spot these minor irritants while there is plenty of time to find elegant solutions. There may even be enough time for a pat on your own back.

> The difference between a problem and panic is measured in the amount of time you have to sort it out.

The invisible project manager

If you were very successful in doing this you would have a very excellent project. It would probably be your last, but your last project would have been a good one. It would be your last because lesser mortals like your managing director would simply dismiss your success. With people saying things such as 'That was easy, what do we need a project manager for?', your career will take a significant nose dive.

In the totally successful project all problems are avoided, the project moves forward to plan and to cost, and all those efficient people on the project simply get on with their jobs. Slight problems are spotted well in advance and are neatly, adroitly avoided. The problems are not even seen by those not immediately involved. Those involved are not even seen by those not immediately involved.

The totally successful project manager is therefore totally invisible.

The company, next time it is faced with a project, may move ahead without a project manager and be quite upset when the project collapses into a gooey mess like one of those pizzas you find on the pavement outside a takeaway. When this happens the senior executives will put it down to bad luck! They will complain that they got a project manager on the easy job and didn't get one for the really hard one.

So what we have to do is one of two things:

1. Totally avoid project management.
2. Dodge problems, finish the projects on time and to budget, and get applause for doing it.

The following pages are aimed at those adopting option 2.

> Those choosing option 1 should throw the book in the nearest paper bank and recycle the paper.

From time to time you will see the TRUMPET icon. This is to indicate that you should think how the preceding section presents an opportunity for you to make your presence known, become visible, and advance your own career prospects.

Project managers are not wizards and do not carry magic wands. There is a common view that project management is the latest whiz and will magically solve all problems. Project management is merely a collection of techniques and attitudes that may bring a little light to a confused commercial world. Unfortunately, there are salespeople driving around in their company cars trying to sell their project management software package. These salespeople will tell you that the package they are offering is the solution to all your woes. The result of this is that:

1. The salesperson gets chucked out for being very silly.
2. The purchaser falls for the line, buys a copy of the program, drops it in some poor, unsuspecting person's lap, and announces that that person is now a project manager. All the budding project manager has to do is install the software and start it running, and profits will instantly increase.

This is, of course, nonsense. Exactly the opposite happens, as the company has just spent some profits on buying the software and will, in the short term, see no benefits at all. Project management, like any other technique, needs to be understood and learned before tools, and that is all a software package can ever be, can be usefully wielded.

Project management is a profession

Once upon a time accountants originated within a specific industry. Now you can be an accountant and know next to nothing about the industry in which you happen to occupy a desk. Accountants, these days, move easily from industry to industry.

Project management has not yet reached this stage. In some industries project management people grow out of those who understand the industry. It clearly helps to understand what the project is all about. Increasingly, however, project management is becoming industry independent. It covers topics and areas outside the scope of an industry and becomes applicable to many. The techniques are applicable regardless of what is being attempted.

You find these project management specialists everywhere. You may have had a sheltered upbringing in a very nice industry, and it may not have occurred to you that people elsewhere use exactly the same techniques and terms as you do.

Project managers want to complete projects on time and to budget. Such people have the responsibility for achieving some set goal and may not realize that they are project managers at all.

Some people are given the title of project manager for no good reason. Schedulers who work for construction companies and engineering firms are obvious candidates. Here are some from an enormous range of people using project management techniques. They often have special problems, problems that are highlighted in their particular industry.

Here are a few examples which will serve to get you thinking about the broader scope of project management.

Information technology

There are a huge number of project management people in both the hardware and software industries. These high tech project managers have to co-ordinate the work of a number of software and hardware engineers, analysts, and programmers to produce a system of some sort. Often certain parts of the package cannot be written or tested until other parts are complete. Installation work and site testing depend on the package being complete and ready for customer acceptance.

There are projects where the objective is to design, write, and launch a project management software package. They get everywhere, these project managers.

Organizational change

There is a vast horde of project managers working to deliver organizational change in both the public and commercial sectors. In any central government department or local authority you will find teams of people struggling to deliver changes that will help to meet their organization's targets. In many ways the future health of the nation is in the hands of the project managers working within the labyrinth of the NHS.

This is also true of the commercial sector where banking, as well as the rest of the financial sector plus many manufacturing organizations, strive to improve their organization's performance through improvement projects.

Construction and engineering

This group covers an enormous range of people. Contractors who are building bridges, hotels, housing, oil rigs, or power stations are typical project managers. Our lives depend on them. They get us under rivers, over mountain passes. They build hospitals, schools, blocks of flats for the rich, and prisons for the unlucky.

Publishing

Here the objective is to get some kind of published work out on time. A good example is a travel firm publishing brochures promoting its package tours. A travel brochure involves negotiating contracts with airlines and hoteliers, getting pretty people to pose for pictures on assorted beaches, writing all that garbage about the flat being ten minutes from the beach, and printing the whole thing in time for the Christmas advertising drive.

Marketing

One of the most common uses of project management techniques is in the launch of a product. Consider for a moment the steps involved in launching a new product, and you will see that this is a sizeable project. You have to design the product and its wrapping; arrange distribution and pricing; make your advertisements; and book space with TV, newspapers, and cinemas. You may have to train a sales-force and paint up a fleet of trucks.

If the product is not ready in the shops when the adverts appear, heads will roll. If the product is too early, with the result that you have your new chocotoothrottobar stored in huge warehouses waiting for the advertising campaign, it's Madame Guillotine time again.

Factory managers

Project management is an ideal tool with which to predict the complex project activities involved in installing a new production facility. All those activities in selecting a new production line or printing machine and getting it installed and working can be planned and tracked.

I once did a little work with a group of people who planned to install a production line making 160-odd packets of breakfast cereal per minute. Who eats all this stuff?

> Running a factory is not a project; building a factory is.

Shutdowns

If you have to organize a plant shutdown, then you are running one of the most aggressive projects possible. Here there is time to preplan and to organize materials and contractors. But once the project begins, time is absolutely everything as many shutdowns run 24 hours a day but last only 2 weeks. Out in the North Sea the oil and gas industry runs short sharp projects like this. Plenty of time to get ready but 24-hour panic once the starting pistol goes.

Engineering design projects

Where a team has to design a new product, a project exists. A good example is a new car. A team of people examine engines, ergonomics, market demand, aerodynamics, and many other aspects. These are often interrelated, and a time scale exists if only because the design team are usually very expensive. The design of a new computer is very similar.

Scientific projects

The development of a bit of scientific equipment is quite complex. Developing such machines requires the co-ordination of expensive engineers, physicists and other scientists, and computer people, and therefore makes a project.

There was a project to develop an electronic sniffer. This box of tricks looks rather like the gadget that is supposed to detect guns at airports and actually detects gold teeth, bangles, and people that like to be searched; a sort of squared off, person-sized archway. As you pass by this sniffer, air is passed over your body and thence through a spinning disc which connects to all sorts of electronic wizardry that is able to detect minute quantities of explosives. One such machine awaits Guy Fawkes's next visit to the Houses of Parliament.

Space

All those bits of hardware being shot into space so that we can watch 19 channels of rubbish on TV are very big projects indeed. Shuttle launches, telescopes in orbit, and moonbase stations are all NASA projects.

There was a guy in Los Angeles whose project was to stitch an extra metre-wide ring around a satellite dish. This dish was listening in to *Voyager* as it zipped around Uranus (stop that laughing at the back of the class) sending back signals. The reason for the need for a bigger dish was that the probe's batteries seemed to be lasting longer than expected and the poor little thing had no one to talk to. Getting a bigger dish is no trivial project, as the degree of accuracy required makes normal engineering look very rough and ready.

Defence

Regrettably, defence is a major user of project management techniques and tools. In defence everything they make gets thrown away. Sometimes, the product gets thrown away inside someone's head; occasionally, but by no means only, inside the head of a member of the enemy. The things thrown away are very high-tech indeed, with all sorts of clever devices that guide and watch and sense heat before they go bang.

The development process is strange and wonderful. One of the various groups that might one day be enemies of the others brings out a new cleverer device. The others employ spies to find out about it or read about it in scientific journals. The others have to find a device to overcome the new one, a process which results in more new devices. This weird roundabout of defence development means that the US can spend huge sums on developing an aeroplane that cannot be detected by radar. This of course will mean that someone will have to invent a new system that can detect the US plane.

Entertainment and sport

If you have ever made a home movie or organized a local sporting event, try to imagine what it must be like to make an epic film or organize the Indy 500 or the Olympic Games. There was a TV programme on Channel Four about the project management of the Calgary Winter Olympics and the Lombard RAC Rally. Both of these are run like good projects.

Organizing a charity event

Things like the London Marathon take an awful lot of getting together, and it takes good project management to bring the whole thing to a successful end.

We could rattle on for ever. There was a project management team working on the artificial insemination of 5000 cows in Malawi. Airlines project manage the major services that their jumbos get from time to time.

The results of a survey of readers of *Project Manager Today* magazine in 1991 revealed the industry breakdown, shown in the illustration on p. 10.

Does the fact that all these people are tackling all these weird projects make you feel any easier? Does it give you the willies? Do you notice some characteristics that all these projects have in common? Here is one thing that all project managers have in common. Everyone in project management knows with great certainty that his or her area of project management is different from everyone else's.

Builders know for sure that building is special because they have the weather to contend with. Designers are equally confident that the unknown nature of their work, the fact that they don't know what they are trying to achieve until they have finished it, makes them special. Network Rail, who can only design things during the working week and can only work on the track on Sundays, sets itself apart on this basis.

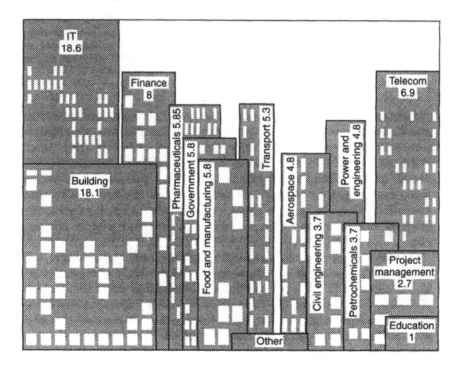

Their misplaced but certain knowledge that they are different is one thing that all project managers share, and they are all wrong.

Getting the words in the right order

In this chapter advice is given in respect of the terminology of the subject, attempting to ensure that when conversation centres on project management, the reader will at least be able to use the correct terms in the correct order. Here the major terms will be explained, taking very little previous experience for granted, so while the chapter may be, in part, superfluous for experts, it will clear the air for beginners.

Project management has become a hot topic over the last few years and there are some very good reasons for that. With many new techniques and, for that matter, technologies, the people who by chance happen to be in the right place at the right time tend to get rich quick. Attracted by this sudden wealth they keep all the (usually) simple ideas hidden underneath a cover woven from abbreviations and jargon so that less fortunate people are kept out.

The theory is that by keeping outsiders out, the insiders will be able to make more for those insiders kept in. Such people are term-droppers. This was particularly noticeable with the sudden proliferation of microcomputers since the 1980s. Those in the know kept those out of the know, out of the know. It became a matter of some shame not to understand megabytes and bit streams. Few people care to admit their own ignorance, especially when there is some know-all sneering down his (it's nearly always a man) nose at you because you don't know how the latest chip responds to Windows XP.

The world of finance has become so good at in-crowd jargon that the TV channels even report on a range of incomprehensible numbers each morning and there are long interviews with smartly dressed people talking about Footsey, 'Knickhay', and someone called Dow Jones, who has an index all of his own. Who cares?

Did the ancient Greeks do this or did Pythagoras share his hypotenuse with anyone who happened to be passing through his bathroom at the time?

Those that know little about anything technical drop names instead of terms. Well, that's what Joan Collins told me. Hence the term, *term-dropper*.

Project management had become rather the same. There are those who drop terms like Earned Value Analysis and Work Breakdown Structure in that insinuating way that prevents elementary questioning and therefore understanding. There are those who deliberately start any discussion about project management half-way through, safe in the knowledge that you will not, until too late, realize that you have missed a number of stages essential to proper understanding.

The general result of all this is bad. You get only part of the story and yet feel that you should know about all of it. Because you don't feel like talking about your lack of understanding, you think that you are the only one who doesn't understand, whereas in reality many others share your doubts. Many people get by as experts simply by knowing the right terms. Their one mistake is to lean on the bar, order another round, and proceed to use all the right terms in completely the wrong order.

Firmly, flying in the face of convention, we are going to start at the top. Rip back the covers, junk the acronyms; we are going to start at the start. You can skip bits if you like. To make life a little easier, each new, vaguely technical term will be introduced in **bold**. If the meaning of the term is a little vague you could consult Chapter 14 for a definition.

Introduction to project management

What is a **Project**? That's a tricky one to start with. The Association for Project Management had a working party on that. If you could find a one-line statement that neatly answered that question, you could be on your way to stardom in project management. Most answers seem to run to half a page and then leave a lot to be desired. Try this brief statement and series of characteristics that most projects seem to have in common:

> A project is a human activity that achieves a clear objective against a time scale.

Projects nearly always have the following characteristics:

- one clear objective;
- a fixed time scale;
- a team of people;
- no practice or rehearsal;
- change.

Let's run through those to get you in the project mood. These are not black-and-white rules; they are trends and you may find that your projects buck the trends.

One clear objective

The classic project has a simple singular objective. Build that pyramid, launch that rocket, construct that hospital. Such projects are simple, clear, and comprehensible. Project management had its roots in power stations, bridges, and things like that. Such megaprojects attracted project management people like flies, and most people worked full-time on one project for a long period. Not very many power stations are built, and thus it was a specialist area.

What has happened over the last decade is that project management has come to the masses.

We shall look at what has changed a little later.

Projects these days are not always so clear and specific. Project management techniques and concepts are applied to a wide range of tasks that really do not have this singularity. The techniques and tools have spread into areas where there is a degree of repetition. An engineering workshop might handle 100 or so jobs each year. Each job involves designing and manufacturing a batch of gearboxes. A software house might have a dozen or so projects on the go at any one time.

We shall discuss the range of projects and how differences shift emphasis from time to resources. It is still true to say that unless you can identify a clear objective, or at least a short list of clear objectives, project management may not be the right tool for you.

A fixed time scale

There is usually some time scale attached to the objective. There is frequently a fixed **end date** that everyone has in mind for each object-ive. The power station opening, the bridge, the horse-race, or the product launch are all planned to happen on a specific date.

It is a fact of project management that you work yourself out of a job. Each project comes to an end, and at the end everyone in it has to look around for something else to do. This creates interesting motiv-ation problems.

There is often a specific starting-point for the project. Sometimes projects seem to emerge from under a gooseberry bush; sometimes there is a definite commitment to the project accompanied by a fanfare of trumpets. Successful projects generally allow some time for **planning** before they get started. In unsuccessful projects, planning happens when things are already going wrong.

Projects, having definable starting- and end-points, are differentiated from continuous processes. Within some companies a project starts when approval is granted by the relevant committee for work to proceed on the scheme and ends when the objectives are achieved, when the **deliverable** is delivered – for example, 'when the road is opened'.

So far, then, we have a clear objective that has to be achieved within a time scale.

A team of people

Projects are human endeavours. No project can be done by machines alone. Excluding beavers building a dam or termites building a mound, projects are done by people.

Of course, project management people do not get their hands dirty. The project management team just organize, manage, plan, and try hard to avoid doing too much real work. Project managers collect the bits, and bits get put together by others. When project management people talk about rolling up their sleeves and getting some work done, they sound very enthusiastic and important. This work nearly always gets done in the conference room and the only muscles that get exercised are those attached to the jawbone.

No practice or rehearsal

Once, in the good old days of project management, this was entirely true. Most projects involved cutting completely new turf. No one had created a Polaris Missile, a Drax power station, or a Eurotunnel before. These days, as project management spreads to more normal projects, there is more experience to draw on. You might have published the catalogue last year and therefore can call on that experience. Building a housing estate gives the builder loads of practice.

Nevertheless the particular project that you are about to undertake is unique. No one has ever done quite what you are about to do. You would be daft to ignore experience gained on similar previous projects, but the combination of time, place, people, and project is certain to be unique. Perhaps the key point here is explained by a factory production line. If the factory manager has a bright idea and thinks he has a better method of canning beans, he can try it out and see if it works. He can watch various factors to see if things are better or not. Using this knowledge he can decide to adopt or discard his new idea.

Such options do not exist in project management. You cannot normally try the project a few different ways. You normally only get one go at it. Because projects are unique, it is very hard to measure how successful a project has been. The answers are always subjective.

People say things like: 'I thought the project went well', or 'Our opinion was that it went badly.' You cannot give a project a factual score, as there is nothing to measure.

Change

The function of many managers is to keep things exactly as they are. A manager is often like a policeman – his job is to maintain the peace. The factory manager's function is to keep production of WazzoCrunch going at the same rate.

Not so for the project manager. When the project is over, the world will be a little different. There will be a new bridge, a new circuit board, a new film. The essence of project management is to create change. Some projects involve removing something. Decommissioning a nuclear power station is a large and tricky project which leaves a field where there was once concrete and steel.

Therefore, the role of the project manager is to rock the boat. If you don't like change, project managementville is not the place for you.

Many projects are just the sort of thing that project management was designed to help with. Projects usually involve combining the efforts of a number of people to achieve some set goal against a time scale.

This perhaps leads us to a definition of project management.

project management is the management of change

This is not the official definition. You will find few textbooks with that definition in them, but it does seem like a good one. It is a little over-simplified but not a bad stab.

The argument uses the concept of **states** just like the states that are used in physics. To manage a project is to manage the movement from one state to another. The change can be perceived by comparing the two states. The end state will be different from the start state and the difference is what you are trying to achieve, what you are trying to bring about.

There was a state where there was no Humber Bridge. Driving out on to the river bank would have resulted in your car getting stuck in six feet of sticky mud. Now there is a bridge there, so that driving out on to the river bank results in your driving across the river and paying a large toll. A change has resulted.

Even if the deliverable of the project is not so permanent, there is change.

If no one sets about organizing this year's National Rally there will be no rally. The project management task is to ensure that a rally happens on the right date and at the right place. The state before was one in which there was not going to be 170-odd cars tearing around the British Isles. The state after the project is one where the drivers have done their tearing about; got tired, wet, and hungry; and two people have showered many others with champagne.

What is not a project? It's worth bringing this little section to an end by painting a picture of what is not a project. What we really mean by this is, where are project management techniques not likely to be of much use?

If the definition works well, things that are not projects will not involve change.

Clearly, any continuous process is unlikely to benefit from project management tools and techniques. That plant churning out millions of gallons of fluorochloroacetylbenzoate, the team collecting the toll on the Tyne Tunnel, and fisherman; such groups don't find themselves with a burning desire to get into project management in a big way. They don't have definable start- and end-points, they don't have clear objectives, they can practise and rehearse; therefore, they don't have projects. They do not want to bring about change.

Not very clear it is? The border between projects and non-projects is fuzzy and tends to get fuzzier still. Where does a house builder fit? Where does a jobbing engineering works fit?

What is project management?

Project management is a collection of loosely connected techniques, some of which are useful in bringing projects to a successful conclusion. Clearly, the project manager manages. He must think about motivation, team building, career growth, financial control, and all the other things that concern managers. In addition he has to head off into the unknown. His path is not clear, his path has not been trod before, but his objective will be clear.

Something that is special about project management, something that separates project management from plain old management, is the need and the emphasis on planning. Simply because the project management team is following no known path, they must think ahead all the time. They are continually faced with decisions about the route ahead and must plan for events that are long distant in the future.

Hence here is a formula that neatly paints a picture:

$$\frac{\textbf{management} + \textbf{planning}}{\textbf{project management}}$$

It is not that simple of course. Nothing is. There are differences in the way managers manage in the project environment which result precisely from the demands of that environment. For example, teams are often flung together at short notice and need to get down to work quickly. Team management requires greater consideration in project management than in normal conditions.

Most managers plan ahead. They order stock in good time; they warn of impending problems. But in project management, planning is the driving force. Behind every successful project there is a successful plan. To get to grips with project management, we have to get to grips with project planning.

What is project planning?

Planning is primarily about thinking ahead. If it is the project management manager's job to manage what is going on, it is the planner's role to worry about the future. Now you may not be lucky enough to have a full-time planner. You may wear two hats – one saying project management on the front and one saying project planner on its peak. Perhaps you have one saying Kiss Me Quick on it. Someone, somewhere, sometime, has the function of being the planner and trying to predict how the project will go and what needs to be organized for the future.

There are three simple objectives in project planning and project management:

50%	Thinking ahead
25%	Communication
25%	Yardstick

Let's take a look at these three objectives.

Objective one: thinking ahead

Half the value of project planning is to provide the opportunity and motivation simply to get people to think ahead about the project they are undertaking. This process tends to reveal problems and therefore helps find solutions to them. Problems get solved while they are still small, remote problems. Few problems are overlooked and left until they loom large. The more people that you get involved in this thinking-ahead process, the better the project will be. You cannot sub-contract this thinking ahead. You cannot get someone else to do it for you.

If nothing else, project planning systems provide a focal point about which the project team can spend some time thinking about the future, spotting problems and overcoming them. Few things can be as useful as thinking ahead. People say, 'I haven't got time to plan; I am too busy writing this specification [/building this bridge/looking for a supplier].'

This roughly translates as: 'I didn't plan earlier and I am so busy overcoming the unforeseen problems that I haven't got the time to plan now.' You can be fairly certain that this will continue until the project ends some considerable time after the planned end date.

The question is close to this one: 'Do you want to go on throwing buckets of water on each fire, or shall we build a fire engine?' So plan, plan, plan, and then plan some more. It is rarely wasted. Do not get someone else to do it for you just because the director/client/project director wants to see a plan. They want to see a plan not solely because they want a pretty picture upon the wall; they want to know that you have planned. Paying lip service to planning can only be useful to you personally; it is unlikely to help the project very much.

There is a major international manufacturer of computers, printers, plotters, and many other computer-related devices who has a name-sake that you put on sausages. When this firm's managers decide to create a new printer, they form a team and get the team members all together for the day in a room with a project management software package displaying **barcharts** and **network diagrams** on a large screen.

They talk the project through. The technical authors explain what they need to know before they can start writing the instruction manual. The hardware team talk about mock-ups, colours, and physical design. The software team get to know when they should be able to start and estimate how long the programming will take. The marketing people get an idea of what they are going to sell and when they can expect pictures of the new device for the brochure. At the end of the day they have a clear plan for the project. They know each other; they know each other's roles. They have an idea of how their work depends on that of others and how others depend on them. They know what to expect.

Brilliant. That doesn't mean to say that all their projects work out; things are bound to go wrong. But what a starting-point. Such planning is expensive, but they feel it a worthwhile investment.

Of course you could organize a meeting. Perhaps not as comprehensive or expensive as the computer firm's, but you could get the key people together to talk about the project and its how, when, where, who, and with what. You might even appear efficient as a result of doing it and gain points by explaining how much has been gained through the planning process.

Objective two: communication

Second in importance as an objective for project planning comes the ability to produce and issue reports to everyone so that the proposed timing, method and strategy are available and understood. Not many companies can get everyone involved in the project together to plan the project. There are usually outsiders who will join the team later. There are suppliers and contractors who are not yet identified. Some people might have been off sick.

But you will be working as part of a team.

Therefore, project management teams produce reports to tell everyone what is going on; more precisely, what is planned to be going on. You want to be sure that people work together. If your next project was a pipeline from the coast to the city, it would be useful if everyone plans to start at the same end. If the trench diggers start at the coast and the pipelayers start at the city, many children will be lost in the gaping chasm left open for many weeks. If we tell everybody their part in the project and how their work affects other people's work it is more likely that everybody will work together in an orderly, organized, and efficient manner.

The **multiproject** or **programme management environment** is very common. In this arrangement a number of simultaneous projects overlap, and team members are not always working full-time on one project. It is important that everyone understands when each project is planned to go through its stages and when **resources** are required for it. You can see that it will be possible to estimate the departmental needs for resources on each project and therefore on all projects in the future and only to undertake projects for which resources will be available. The need for communication in a multiproject environment is even greater.

See Chapter 12

Don't forget your senior management. They do need something to stick on the wall to show how clever they are. They picked you, didn't they? Also, you, perhaps, have cleverly included all sorts of tasks that other people have to do on these reports so that the ball is normally in someone else's court. You may have action items for your own boss. If these people don't perform, you can, at least as a last resort, claim that you told them what was required.

You can discuss the information requirements with your senior managers and provide them with exactly the format they requested. You can check from time to time that they are getting the information they need. The wise project management expert chooses to ask such questions when the project is going well and keeps out of sight during other times.

Objective three: a yardstick

'Things often do not go to plan – what is the point of planning?' The plan also provides a yardstick against which to monitor the project. At the least, you can measure how well the project is going.

The plan is very much like a budget. Let's imagine that you have decided to build a house for yourself, your spouse and your everloving kids. You decide on a few basics like the overall size and number of rooms and prepare a budget. The budget lists the items on which you expect to spend money and indicates how much you expect to spend on each.

On the basis of a satisfactory budget you commit to the project. Each time you spend some money you check it against the budget. If you find that the roof costs more than you planned, you might choose cheaper floor coverings. All the time you are in touch with expenditure and are able to make sensible decisions based on the state of the budget to date. You do not expect to spend money exactly as the budget says you should, but you have a model to monitor expenditure against.

A project plan is like a time budget. You allocate time to various tasks and use this as a model against which you compare the actual use of time. If something takes longer than you think, you try to find a quicker way of doing something later on.

The slight difference between time and money budgets is in the way you add them up. In the budget every pound must come from the family coffers; therefore you simply total up all the items. In time, many tasks can go on simultaneously, and thus you need a different technique – adding up the times will not do.

Projects never ever go according to plan. There are always deviations, hesitations, and interruptions. There are a million things waiting out there to make your project go wrong. The engineer's girlfriend gets pregnant. The financial director's boyfriend runs off. Things arrive late, are to the wrong specification, and simply don't work. There are earthquakes, tidal waves, previously undiscovered archaeological remains, and a protest movement.

Many things can go wrong. You can't stop them. Murphy's Law will be obeyed – anything that can go wrong, will go wrong.

There is another problem. We human beings are poor at predicting the future. Nostradamus and others have tried. Various religious sects predict the end of the world. They are generally wrong, so far. Life would be deadly boring if we knew the future.

Looking into the future is like looking into the distance. Far-away objects tend to be unclear and often partly hidden by intervening hills and other obstructions. The further into the future we look, the less clear it gets. Hence projects seldom go the way they were planned.

One intelligent project manager said that if you drew up an infinite number of plans, the actual work would always follow yet another one, one you didn't draw.

It follows that plans are neither wrong nor right. Apart from any obvious flaws (like planning to put the roof on before the walls are built – for which you should be stood up against a wall and pelted with old conference table ashtrays), plans just are. Put any two capable planners on to a project and they will come up with three workable plans and nine workable excuses.

Regrettably, we cannot control the way things go wrong. This is not a reason not to plan. It is a good reason to plan. We do not attempt to accurately predict the future. We model the future and use the model as a basis against which to **measure progress**.

Therefore face up to the facts – the project will not go as planned. Now let's see how we handle that. What we must and can know is how things are going. We need a yardstick against which we can monitor progress on the project. If we don't have a yardstick, we have no means of knowing what is happening. We cannot even say with any confidence that the project is ahead of, on, or behind schedule. We need our time budget.

Thus, the project plan becomes our yardstick by which we can regularly and frequently monitor progress. A plan should not become a stick that your boss can use to beat you around the ear with. Yardsticks are to let you know about problems in time for solutions to be found. If your boss is the type to beat you around the ear, he or she should find his or her own stick. The boss should thank you for having the intelligence to read this stuff, plan the project properly, and go to him or her early with the problems.

You could establish a regular project review meeting at which the current status of the project is discussed and future actions chosen. You could chair these meetings and, assuming that people think that things are going well, reap praise. Even if things are not going well, you can point out that good planning has made sure that everyone knows about the problems in time to overcome them.

> It is not necessarily a crime to be running late, but it is always a crime not to know that you are running late.

How do you project plan?

Project planning is a modelling process. Project managers create a model of the project and experiment with the model to find neat, efficient, and cheap ways of proceeding through the **activities** or **tasks** which make up the project.

Simple projects can be adequately modelled by a barchart – this shows the activities which make up the projects and their timing. It does not show why they are planned to occur at that time. More complex projects often are helped by the use of a more complex model.

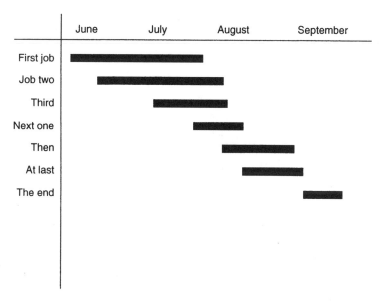

A barchart.

This more complex modelling technique allows the analysis of what is going on in the project and also allows the project manager a convenient way of telling other people what the plans look like. The model is sometimes called a **PERTchart,** a **network,** or a **critical path diagram**. These models are the input side of the process; they define the activities and how they relate to each other. (See figure on p. 24.)

The output side is the 'when'. When should activities start and finish? A barchart is one of the reports produced most frequently. It shows the activities represented as bars drawn against a time scale. Barcharts (or Gantt charts) communicate the plan very clearly.

The analysis produces information detailing when each activity should start and finish. From this information you can draw up a barchart with great confidence. Also the calculation identifies which activities are vital to the timing of the overall project **(critical activities)** and which are less important.

PERT is a route to a good barchart.

Additionally, project managers calculate how many resources an activity will need and therefore calculate how many resources will be needed day by day. A graph showing the amount of a resource that you think you will need is called a **histogram**. (See figure on p. 25.)

By distributing barcharts and histograms to the people involved in the project and the people involved in your future projects, you can make everyone realize how much work you are putting into the project.

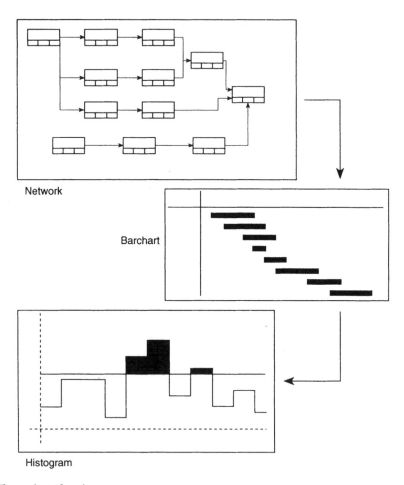

Network

Barchart

Histogram

The project planning process.

Measuring progress

Measuring lateness is an art in itself. Let's talk briefly about measuring lateness. You often hear project managers saying things such as 'We are two weeks late.' This seems to be quite meaningless. If someone makes a statement like that, you should ask questions such as: 'Which part is two weeks late?' 'Two weeks later than what?' 'How late are the critical activities?' A phrase such as 'two weeks late' in itself is pretty meaningless. It is as useful as '2 metres deeper', '1 millimetre longer', or '1 mile higher'. Such phrases are useless without some basis for comparison.

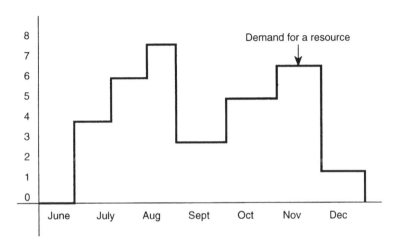

A histogram.

If you really want to know how late the project is, or really want to tell someone else about the project, the only true measure is to compare the currently predicted project end date with the project end date as it was predicted some time ago. Thus, a statement such as, 'We currently expect to achieve project completion on 17 September 1996, which is 5 days later than the target plan', is something you can say with confidence. This is something everyone can relate to. If you want to be honest this is the way to talk about lateness.

If you don't want to be honest, you can revert to the 'three weeks ahead' approach while you pack your suitcase and search the classified job advertisements in the back pages of your professional journal. 'The honest approach' works equally well using intermediary **target dates** – you could set up a target date, for example, for 'Committee Approval' and monitor progress against that date.

Updating

A project plan is a model, a time budget. If a project plan is a model of the project, then by using the model the project manager can perform another useful service – **updating the plan**. This is where the computer will really help you.

The reason for updates is that projects don't stay still – they constantly change. As work gets done the amount of work that remains to be completed reduces. Therefore the project is changing by the day. It is getting smaller. As you achieve parts of your project, there is less work remaining to be done and therefore the project is getting smaller.

> You can only manage the work that remains to be done.

Consider for a moment a house – you probably live in one so the concept should not be alien. At the beginning there is the whole thing to build: foundations, walls, roof, and then the interior. Then one of those yellow diggers turns up and hacks away at the ground. After a few weeks' work, the foundations are completed, and then the project involves walls, roof, and interior. A few weeks later the walls are up and the roof is on. Now all that remains is the interior work. Finally all that remains are the light-bulbs.

Thus, as time goes on and work on the project gets done, the project gets smaller. The amount of work that remains to be done reduces. It may not reduce in the way you predicted but the chances are it will get smaller. It's a bit like pushing a ball of Plasticine up a hill. Occasionally bits bulge out and need collecting in again, occasionally bits fall right off and need to be picked up and stuck back in. The general direction is, however, onward and upward.

While we are drawing analogies, another one is to say that a project is like herding a flock of sheep along a road. The sheep are the tasks. Some wander off and need bringing back into line, some lag behind, some get ahead of the flock. The shepherd watches what is happening and takes action to collect the errant sheep back into the flock and to keep the flock heading towards market. The shepherd whistles and the dog does all the hard work. Almost exactly like project management.

> Project management is like herding cats.

We can use the project plan as a model which gives us a basis for a regular review process. At these reviews we can:

- Look at what has happened since the last review.
- Plan how we will proceed from here.
- Prepare a revised model of the rest of the project.
- Distribute information communicating this plan.

Thus, the project plan gives you a yardstick against which you can measure progress and a basis for a regular review and update process.

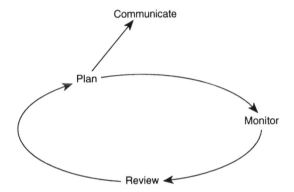

In this way we get a regular plan/monitor/review cycle which keeps planners employed and projects up to date. On such projects a new project plan is put up on the wall once a week to let everyone know what is going on.

Nine steps to a successful project

This chapter takes our hero, the project manager, through a series of steps essential to a good project, and comes out shining at the end. The assiduous reader who takes the time to study this short chapter will gain an insight into the true nature of a successful project and will become acquainted with many potential and easily avoided errors.

It can be done. You can get a project completed on time and on budget. A successful project is one that finished on time and on budget, and that met the quality standards. And it is one on which no one suffered a heart attack caused by strain.

Let's assume that you have been assigned to be the project manager on a new project which seems likely to get approval to proceed soon. How are you going to tackle this seemingly daunting task? We shall look at what a successful project manager does before a project gets going and during the project itself, and at how projects are planned.

The nine steps

Step 1 – define the project

It is vital that you define what the project is about, and how much of it you are responsible for. This might be laid out in a specification and some drawings. It is very important to know which parts you are doing and which bits someone else is responsible for. At a minimum you should get a clearer definition of the deliverable of the project.

If your organization uses a formal project management methodology your project might be defined in a Project Initiation Document (PID). Chapter 9 explains more about methodologies.

Then you can break the plan down into activities and start to think about how the activities depend on each other. At this stage the number of activities can be very small. The first attempt might only have six or eight activities representing the major stages in the project. This will normally have been done at an early stage. You can break these down into more detail later.

Step 2 – build the plan

Build the plan a bit at a time. The second attempt might involve only 20 to 30 activities and provide a summary of the plan. At this stage you should be able to get a feel for the overall, broad-brush timing of the project.

The first shot often works out to be too long and too expensive. Do your best to be realistic and be prepared to justify your plan. Remember that the whole project might be cancelled if it looks like it will take too long. Eventually, after some discussion, you will get an overall plan agreed with all concerned. Next you can start to build detail into these first few activities. As the plan gains more detail, it grows in size, complexity, and accuracy.

There are advantages in clearly stating target dates and in choosing auspicious dates. 'Getting the roof on by Christmas' is the sort of thing people might be motivated by.

Step 3 – get the plan prepared

This part depends mostly on the system you are using. It might mean getting a neat barchart drawn in the drawing office. It may mean running a PERT chart through a computer to produce a barchart. It might mean using a planner on a contract basis. It nearly always ends with a barchart of some kind. The barchart is normally your primary means of communication.

Bit by bit, the plan should be developing in detail.

Step 4 – agree the plan with your managers

This involves running through the plan with the senior managers within the organization and explaining why activities develop in the way they do. This will almost certainly involve showing them the barchart, which shows the activities making up the entire project against a time scale, and frequently involves you in a discussion of why and how you intend to execute the various activities. You might show off your network diagram, indicating the logic you have used.

On badly planned projects this stage can often involve considerable negotiation and alteration to the plan. It is likely that you will have lots of changes to make to the plan and that the managers do not really understand the plan. The managers will not feel committed to the plan if they are left in the dark until this stage is reached.

To avoid too much chopping and changing at this stage, involve the managers so that this approval stage becomes a formality. Involve the people who are responsible for executing the project at this stage. Talk through the project or at least the relevant part of it with each manager, each engineer, and try to make sure that both understand the processes involved. This will endear you and your plan to them, reduce the number of missing activities, and generally result in a better plan.

Successful projects have motivated teams running them, and motivation can be created by setting realistic targets that the project team believe in.

If you work away in a corner and create the world's most wonderful critical path diagram, you will have much difficulty in selling your ideas to the others on the project. The reason is that the plan is *yours* and not *theirs*. Try spending time with the people involved getting them to tell you how things will work. You are like a scribe. In ancient days most people could not write; therefore they went to a scribe who translated their spoken thoughts into the written form. The scribe would suggest, enquire, and counsel, and then write down what he thought the customer wanted to say.

Planning is a lot like that. You talk to the experts about the way they think the project will develop, who will do each task and what they will need to help them do it. Then you translate the information into the language of critical path analysis and take the plan back to them to check the ideas.

People tend to be optimistic about planning. If you ask an engineer how long it will take, say, to build the gearbox, he might say 'two weeks'. Then ask him what is involved; perhaps the two of you might draw up a small barchart for the gearbox build. Talk about other work he is likely to have on at that time and don't forget weekends and holidays. By the time you have finished adding realism, you could be easily up to three weeks and on the way to four.

It's a great deal better to know now than later.

By a combination of these techniques the plan belongs to the whole team. They feel a part of the plan, they feel warm about project planning, they feel warm about your role. It is easy to create plans that are used to get back at people: 'You said it would take ten days – it's all your fault' is not an attitude likely to breed mutual respect and a strong team. A statement such as 'We estimated ten days but it actually took longer; we had very little experience to go on' is likely to generate a better atmosphere on the project.

It is a fine balance. There are times when to save your own face you have to blame others.

If you can achieve this positive attitude you are half-way to a successful project already!

Step 5 – disseminate and communicate

The time has come for you to distribute copies of the agreed plan to all interested parties. You could print off complete barcharts showing the timing of the entire project; draw huge PERT diagrams showing the reason for the time; draw histograms showing the demand for every resource. You could then get a small truck and send the complete package to each person on your mailing list.

This approach is very likely to lead to heavily laden desks; desks with ancient barcharts covered in coffee stains hiding in the depths of in-trays. It is not likely to lead to successful projects.

Go to each member of the project team and ask them what they want from you. Then give them the data they want. If the buyer wants a barchart showing only the procurement activities – give him or her that. If the builder is interested only in getting the structure erected, get a suitable barchart for him. This can be achieved by the use of selection criteria. Too much information is worse than too little. At least people complain when they get too little. Too much information is usually greeted with a gruff remark and eyebrows raised to the heavens.

This process of 'designer barcharts' will improve the attitude towards you and your plans and continue the increasing likelihood of the project's being highly successful. Of course there are standard reports that you can generate easily on a regular basis.

As a part of this communication process, you might organize a project meeting, run through the plans, and explain to everyone why things should happen when you say, and why you planned it that way. This might provoke discussion on the process. You might learn a lot about the project, and you will be getting people to think ahead.

> Getting people to think ahead is 50% of the benefit of project planning.

At such a project meeting you should ask a question which is vital at this stage: 'What have we forgotten?' This might promote some more useful forward thinking.

Step 6 – do some work

All this planning is less than useful unless some work gets done – cut some logs, lay some bricks, write some words. Never forget that planning is not an end itself – it is a means to an end and that end is a successfully completed project, on time and to budget.

Project management teams often do very little actual productive work on the project. The rest of the time they do nothing at all. The project manager gets the resources there – the project manager provides the environment where the work can get done. You get the pipes, welding rig, welding rods, and welders all there at the right time, and hey presto! the pipes get welded up. Project managers are facilitators, not doers. Project managers take the horse to the water;

they check that the water is OK and that the horse is qualified to drink. The wise project manager monitors the horse's drinking. But it is the horse that does the drinking.

Step 7 – updates

Now that you have a plan agreed and accepted by the members of the team as a central focal point about which they will all think ahead about the project, you can ask them to help you to update the plan.

Wait a week or perhaps two, and then go on an inspection tour one morning to see how things are going. If you do this yourself, take a copy of the barchart with you and mark up what's been going on since you did the report last time. You might have to ring your suppliers to find out what has been happening. You might check the design office's progress so far. You might design a form that people fill in to tell you what is going on. Such a form might list current tasks and ask for the

---

NORFOLK AND GOODE plc

PROGRESS AND MONITOR REPORT FORM

Project: **Date of Report:**
Prepared by:

Task Number	Task Description	Planned Start I Finish	Actual Start I Finish	Expected Duration

Typical progress monitor form

remaining duration and any comments to be added alongside the tasks.

You need to check only what remains to be done.

Once you have established the status of current activities and any changes that need to be made to the future plan, you can modify the plan to include these changes. Then you can get a fresh set of reports and distribute them.

On many unsuccessful projects the team occasionally meet in the toilets and corridors. On happy projects there is a regular project review meeting. Successful project managers go to regular project meetings, hand out the new plans, and listen to or lead the ensuing discussion. If you have been following all these steps, everyone leaves these project meetings with a feeling of confidence. Confidence in you and your plans and your own reputation is continuing to grow. Things look good. People look forward to their regular barchart as a status report, as a check on progress, and to help them plan their work.

Step 8 – repeat Step 7 as required

Try to continue the update process on a regular basis. You may do this every month. Some project managers do it more frequently than that. Let your judgement dictate the best frequency. Complex, rapidly changing, fast-track projects need more updating than do more relaxed, long-term projects.

If you meet and stare at each other with nothing to say for half an hour, give up and delay the next meeting. Never have a meeting unless there is business to talk about.

Step 9 – receive your congratulations

These congratulations are justly due to you at the end of the project as the new road opens on time and to budget. Try to keep the smile off your face as you get the new company car, the pay rise, and increased respect from your colleagues.

Beware that you do not become invisible – don't forget to blow your own trumpet.

Even if the project is late and has overrun the budget, no one has been surprised by the problems. You have done everything in your power to bring the problems to light and to find solutions to them. Perhaps no sensible solutions could be found. Perhaps your solutions were beyond the scope of the project – the costs of the solution outweighed the benefits.

No one wasted huge sums of money rushing ahead with tasks that need not have been done. On one quite small restaurant project, the project manager was just about to spend a lot of money ordering a very special prefabricated bar from a joinery workshop whose staff were to work overtime to deliver it on time. Just in time, the team discovered a problem with the lease which meant that the bar would not be required for some time. If the team had gone ahead, they would have paid extra for the rapid bar manufacture and then have had to pay extra for storing the thing. As it was, the bar was made normally and cheaply and was delivered in good time.

It is generally better to know the truth as early as possible. It reduces the likelihood of piling wasted expense on top of wasted expense.

The scope of the project and its 4 objectives

In this chapter some valuable lessons are drawn which may be of assistance to the intrepid project manager when faced with a new project. Methods of defining the project's operating parameters and methods of modifying these parameters are outlined.

We will commit to sending a man to the surface of the moon and bringing him safely back to the earth before the end of the decade.

(President J.F. Kennedy announcing the US moon-landing programme)

Defining the project

As we said while talking about the nine steps to a successful project, it is always essential to define the project you are about to tackle. It is important to define the content of the project in terms of deliverables as precisely as possible. As a result, firstly, you will get a precise picture of the work to be done; secondly, you will know the yardstick against which your success will be measured; and, thirdly, you will be able to recognize changes in the scope of the works.

You really need to understand and agree **the measures of success**. Only if you agree how your performance will be measured will you have a fair chance of succeeding. Defining and agreeing measures of success for both you and the team will make life a lot clearer, simpler and fairer.

If you do not have a clear definition of the deliverables and the measures of success, your chances of failing increase hugely.

If you don't believe this to be true consider this argument: your project will have many stakeholders, people who have an interest in whatever you are trying to do. It is very likely that many of the stakeholders will have different expectations. Therefore the results of your efforts may please some people but others will be disappointed; not because you did a poor job, simply because their expectations were different from your own. You will still get the blame.

You can please some of the people all of the time and all of the people some of the time. You cannot please all of the people all of the time. Make sure you please the important people.

If your organization uses a formal project management methodology your project might be defined in a Project Initiation Document (PID). A PID contains a formal way of defining a project including deliverables, risks, benefits, the way the project is to be managed and resourced, quality standards and many other aspects. Chapter 9 explains more about methodologies.

For the moment we will consider a simple project definition form made up of two sections.

The first section deals with the identification of the company, the department, and the personnel that have been involved in preparing the form. If the system within the company requires an authorization, then this too can be included in the first section. Section two deals with the project itself. It needs to be given a name or a number to help people refer to it, and its objective should be clearly stated in a few words. The objective is normally defined in terms of the deliverable (a new building, software package, or report), but it might be in terms of a less tangible objective (move the company to new offices). The objective explains the change that the project should bring about.

Sometimes the best way to define the project is to establish the main elements that make it up – the subprojects within the whole. A precise picture of the deliverable of the project will help clarify your and everyone else's idea of the nature of the project. It forces everyone to be precise. Timing is obviously important. At this early stage in a project's life, the timing will be very unclear but overall target start and finish dates may be set. Cost is usually an important factor, and therefore a section indicating the target budget is usually a good thing. Quality is a very hard thing to define but some companies use standards which do set a quality standard.

A great danger is where the cost and time limits are precise and the project is not. You will find the scope of the project growing, putting you under more and more pressure as the cost and time limits do not change. Often these limits come from different places – the finance director sets the cost target but the managing director wants the highest quality. You become a ping-pong ball batted between the two, and you can only lose.

Get the definition of the project down on paper so that you can be fairly judged even if the people in senior management change.

Changes to the project

Project managers often find themselves taking on extra work by accepting modifications to the scope of the work. A clear definition of the project will reduce this danger. Wise project management teams not only define the project; they also design the system for modifying it. It is amazing how many professional project directors move the project's goalposts frequently. Moved goalposts are a common cause of project failure.

The poor project manager is busy trying to balance the three objectives of the project – time, cost, and quality – when the specification suddenly changes. Someone somewhere decides that the project needs an extra milling machine, an extra floor, a rooftop conference facility, and a helipad. While these may all be vital to the continued success of the company, let's make sure that the project management team have a little extra cash to pay for these afterthoughts and some extra time to carry out the work.

The three objectives of the project are in fine balance, and everyone must be clear about changes to the specification of the project and changes to the deliverable.

In organized projects there is a formal system for changing the scope of the project; this involves **change orders** or **variation orders**. Ideally, each variation order should be submitted as a request in this way: 'We are thinking about adding a helipad – what are the implications?' The project management team then think about the possible change, work out the extra time and costs, and respond. The decision is made to vary the work or not, and everyone is happy – well, at least as happy as they were before, which may not be very happy at all. But at least the decision about the helipad was not taken in ignorance.

Many companies have preprinted forms which are used to request and detail changes to a project's definition. This is relatively easy to organize where there is more than one company involved. In such a case there is usually a contract drawn up between the client (i.e. whoever is going to use the deliverable when it is finished) and the contractor (the company trying to deliver the deliverable).

It is much harder within one company. If you have been promoted to project manager within a company and you are supposed to organize a stand at an exhibition or an office move, then there is not likely to be a contract at all. These are the people who get caught out when the scope of the work changes. You might let your managing director get away with a change, but your painting contractor won't let you get away with any significant change without compensation, usually in the form of banknotes.

Some companies elect a 'client' and a 'contractor' within the same company to emulate the nasty outside world. This has the effect of creating or simulating a contractual relationship between the two parties involved, even though they work within the same company.

Therefore define the project's deliverable as clearly as possible and let everyone know what you think you are supposed to be doing.

Establishing key dates

People often find it difficult to be motivated by far-distant targets, especially if their contribution is short term. For example, it is a weak motivator to tell a foundation worker who has 2 months' work to do that the project must be open in 6 years' time. Also, it is hard to measure progress towards a far distant goal – the rate of progress is often minimal. It sometimes seems that the project is only inching forward towards a target many miles away.

One way of approaching this problem is to set intermediate target dates – often called **key dates** or **milestones**. These are set within the scope of the overall project and indicate significant moments in the project's life. Such significant moments can normally be located – a major approval stage is a typical example of a key date.

Often the nature of the project and the type of resource required change at key dates. Perhaps the project passes from the design into execution phase at a key date. In software development projects there is very little with which to measure progress – you can count bricks in walls and pipes in refineries, but what can you count in a software development project? For this reason software development is broken down into discrete stages separated by key dates or milestones. It gives the project management team something to measure against.

As many software projects are very similar to each other, standard sets of milestones and phases have evolved called **methodologies**. These methodologies explain what each stage in the project should achieve, what information is needed to begin the stage, and what information should be available at the end. Stages have names such as **design**, **coding**, and **testing**.

Key dates aid the preparation of a summary report – the report can summarize the project in ten or a dozen key dates. You can report when the key dates are planned to occur, when they did happen, or when you think that they will happen. Where a number of projects are being run and where these projects are similar in nature, target dates can be standardized, so that any project's overall plan can be quickly evaluated.

Press home these key dates. Let everyone know what they are and how long there is to the next one. There was a large sporting event where the project manager hired a bus and, in the little window at the front of the bus, showed the number of days to go before the big key date – the start of the event. You can erect a flip chart in the office foyer with a little sign saying 'Days to go', and mark the days left on a sheet under it.

There is at least one company that sets an end-of-project party budget right at the start of the project. If the project goes late or over-runs the budget, something is cut off the party budget. The project management team gets to spend the party budget in whatever way they like, and thus a good party might mean a day out hot-air ballooning, a poor party might mean half a cider.

Such simple, little and cheap tricks help to motivate people.

One thing that can easily demotivate people is failure to believe in your plan. If the key dates are unrealistic, the staff will rarely tell you, but they will quietly give up before the project has even got underway.

> Beware people who accept your plan too readily.

Similarly, if there is too much time before the key date, personnel will put off starting on your work as other more urgent matters fight for their attention. Things such as golf, doing the pools, and getting a hair-cut become a higher priority, as staff know that they can start on your tasks at any time. What can easily happen next is that the latest possible start date comes and goes unnoticed, and everything is left to the last minute. Then, and only then, do the last-minute panics develop.

Tread that fine line – find key dates that fall in that range between the date that is too early to be believed and the date that is too late to matter. It also helps to choose memorable dates for your major targets. Within a fairly wide range no planner can predict a specific date; therefore, if you can work your major target to the first day of the month, Easter Monday, Bonfire Night, or some other notable date, you will find it sticks in people's minds.

| Great | Sayings | from | Project | Management | #42 |

A LACK OF PLANNING ON YOUR BEHALF
DOES NOT JUSTIFY A PANIC ON MINE

Evaluation of potential risks

One way in which projects change is as a result of unknown factors. You do not control the unknown; you don't even know what is unknown. If you did, it wouldn't be unknown, would it? You do know the areas of uncertainty, and these should form a part of a good project plan.

Recognize risks as far as possible. If the local authority might not grant planning permission for the new helipad, recognize this at the outset. After all, your company is going to undertake the project on the basis of your proposal or your evaluation of someone else's proposal, and thus the company had better know about the risks. If only you know about the risks, you may carry them yourself all the way to the local unemployment office.

At the beginning of the project you cannot be invisible. Quietly and with confidence, you must make your presence felt and make sure everyone knows what the project involves.

As a part of the project definition, or shortly after the definition stage, there is value in evaluating risks. This process tends to concentrate the mind on those areas where risks are high. Try to understand what investment and work might be at risk if the project is abandoned, and what delays and overspending might result if all those bad things come to pass.

A project is a high risk project if a high proportion of activities have little or no **float**. Critical activities, if delayed, delay the project, whereas activities with float can be delayed without direct effect. Hence, if many activities are on or near to the critical path, the project is likely to be a high-risk project. Within the construction industry, for example, this is much more of a problem in refurbishment projects than in new construction, as you find out what has to be done only when you start stripping out the building.

There are significant risks in research-and-development projects – no one cay say how long it will take to find a cure for cancer or AIDS; no one can predict whether the new circuit-board concept will work at all. The Advanced Passenger Train failed completely because of design problems. Nevertheless, a statement of risks can be a valuable tool for the project manager and the project team. (Chapter 13 looks at risk management in more depth.)

The cost/time/quality triangle

Whatever your project is, it probably has a mixture of the following three objectives:

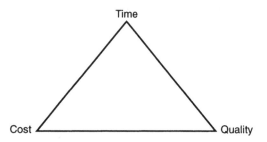

There is value in knowing in what proportion these are mixed in your current job. They pull like vectors on each project. In most projects the total of all three is a constant. If you could find a way of adding them up, you would get the same answer every time. Increase the quality at the expense of expense. Gold-plated bath taps cost more than stainless steel. Taking less time usually means a fast-track project and this usually means overtime – hence extra costs.

Of course when the managing director says he wants to finish the project earlier, he or she will not be happy if you riposte 'Certainly, do I let quality or cost suffer?' 'Neither' will be the unrealistic reply.

We must define quality before moving forward in this area. Quality, in project management terms, does not mean good, very nice, high-class, or lousy. It means 'in accordance with the requirement'.

That's what you are supposed to do – execute the project to meet the requirements. No more, no less. Some people replace the word 'quality' with 'specification', with the result that they worry about cost, time, and specification.

Now let's look at some projects in these terms. Come with me on a magical mystery tour of projects large and small, here and there. Major sporting occasions like the Olympics, Ascot, and the World Tiddly-Wink-Swallowing Championships simply must happen as planned. Short of a major disaster, such events simply cannot be delayed. You cannot report that because of a few problems the games are going to be a week or two late. Thousands of athletes, judges, VIPs, etc., cannot be delayed. Not to mention all those tiddly-wink swallowers.

If there is a problem near the great day, something else other than time will have to be sacrificed – money or quality. If things are getting late the project managers will spend lavishly to get the show going. If money doesn't work (or if enough cannot be found), then quality has to slip. If, for example, for some reason the event programme has not been printed, the first reaction will be to get a printer to work day and night to get it done. If this fails, someone will photocopy the first few days of the programme, maybe in handwriting. Anything is better than nothing.

Let's now look at a motorway. Here quality is rigidly defined – there are specifications to define the required quality standard. Quality is set down in detail. You may find this difficult to accept as you sit staring at coned-off lines of carriageway being mended, but it is true. The specification rigidly defines the quality required. If the end date is approaching, the contractor might report that the project is three weeks late. This will cost him some lateness penalties, and there might be embarrassment, but the road can be late. Official openings of roads and hospitals are laid on months after the contractor has removed the last wheelbarrow and stepladder.

As projects progress, the emphasis moves; once end dates are set and contracts signed, time can become more important than before.

Often different people involved with the same project get pulled in different directions. The client wants a factory to be cheap, the architects wants good quality, and the building contractor wants to do it as quickly as possible. Here is a quotation from a great but hypothetical architect: 'The bitterness of poor quality remains long after the sweetness of meeting the schedule has been forgotten.' How true, but you can see which corner of the cost/time/quality triangle that architect chose to stand by.

Where in the triangle are your projects? How do the three opposing forces pull on you – which are strong and which are weak? It is worth finding out and explaining your findings in the definition of the project. It is worth thinking about the motivation of other people and organizations in these terms.

Who would think there was so much to defining a project?

The spectrum of projects

Some projects are highly logic intensive, and many are not. It is valuable to try to see your project in these terms. When I say highly logic intensive, I mean in the critical path sense. Houses are fairly logic intensive. You cannot erect the walls until the foundations are concreted. You cannot erect the roof until the walls are there to hold it up. Particularly in the early stages a house is a highly logical project. When building a house the sequence is set by the design; work must follow a set pattern – foundations, walls, roof, etc.

A software project is often just the opposite. There is perhaps the basic logic of: **specify, code**, and **test**. But the 17 or so different programs could be written in any order. There is no logic connecting them. What is important is to keep the expensive programmers busy. These projects are resource-intensive projects.

If you had to construct 100 or so scanning electron microscopes each year for 100 different clients, you would know there is no logic connecting the different projects – a delay in one need not cause a delay in another. Where logic is secondary the sequence of doing work is controlled by the allocation of resources – the project manager becomes a resource manager.

It is my belief that all projects fall somewhere on a spectrum. At one end of the project spectrum is the highly logical, stepwise project. At the other end is the intensely resource-oriented type of work. Most project-planning systems are aimed at the logic-intensive project; it's the basis of critical path. The resource elements in project management software are at least as useful as the critical path parts of the software in resource-intensive projects.

Another point I have noted is that in the logic-intensive projects the manager usually is concerned with minimizing the demand for resources as the project follows the required sequence. The manager tends to be involved with one-off projects. If he or she can get by with fewer resources, fewer staff are recruited and the manager feels pleased.

In the resource-intensive projects the project manager tends to be running a group of people – he or she is interested in maximizing resource utilization. In other words, there is a fixed labour force, and the manager wants to keep them busy. Such managers employ highly skilled engineers, who cost a lot of money, and the work must be scheduled to keep them fully occupied. That's what project management involves in many companies.

So where on this scale is your project? Highly logic intensive? Highly resource intensive? It is useful to know, and the answer can seriously affect the choice of tools that you use. Let's summarize the points in a table:

Logic-driven projects	Resource-driven projects
Important critical path	Critical path not so important
Very logical process	Little logic involved
Resources may be low priority	Resources are first priority
Few projects	Many simultaneous projects
Requirement to minimize – resource requirement to be kept low	Requirement to maximize – resource utilization to be kept high

The need for planning

To end this section on the definition of the project, let's turn to project planning. This is the topic we are going to discuss next. Planning means looking into the future and trying to predict how things should go. When do you think the greatest need for planning exists? Whoever answered 'Right at the start of the project' can go to the top of the class. And when do you think we are at the stage when we know least about the project? Yes, the same answer.

It is a sad fact that we need to do as much planning as possible at the stage in the project when we know least about it. Here we are just beginning to define the project, and everyone wants to see a plan. Try and make others see problems you are facing.

The following graph shows how the need for planning input varies over the life of a project, and the graph tries to divide the life of a project into some sensible stages:

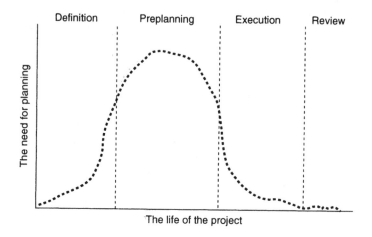

Project planning

In this chapter the intricacies of project planning, network analysis, activity on arrow and precedence diagrams, and similar pursuits are outlined. Assorted clever tricks are outlined for people about to plan a project, and some guidance is given on the selection of tasks.

If you have bothered to read Chapter 1, 2, 3, or 4, or perhaps all four, you will have begun to realize that project planning is a pretty important technique in project management. We had better get you to a stage where you are at least a better planner than those around you, in order that you can do one of two things:

1. plan projects successfully;
2. pretend that you can plan projects successfully.

Let's first of all take a look at the two major types of planning documents.

It is very likely indeed that you will arrive at a barchart eventually. As Caesar most likely did not say, 'All roads lead to the barchart'. The barchart is the single best way of showing what is planned to happen. If you can conceive of a better form than a barchart, please contact me first – I'll make us both very rich. The idea of a barchart is credited to one Monsieur Henri Gantt, a Frenchman, who is famous for two things. One is inventing the barchart, and the second is for having two Ts in his name. Hence, the terms barchart and Gantt chart are interchangeable. They mean the same thing.

Critical path analysis (CPA) was not invented by Monsieur Critical and his friend Monsieur Path. No one seems to have a very clear idea where it came from. Certainly, it sprang into prominence on the Polaris Missile Project, which was undertaken by the US Navy in the late 1960s. Rumour has it that the technique actually came from Germany, but that is just *Volkssprechen*.

While a barchart is a bit of paper representing the project, CPA is a technique which results often in a drawing representing the project. The drawing is often called a network diagram. These days critical path analysis, PERT, critical path networking, networking, network diagrams, critical path method, CPM, PERTcharts, and network analysis all mean the same thing; at least the same basic technique.

Classically, these names meant different things, but who reads the classics these days?

We shall use the term PERTchart.

> We shall refer to these specific techniques later on.

Let's compare the results of barcharts and network diagrams to see where they fit into the greater scheme of things.

The barchart versus network analysis

Barcharts can play a useful role but are limited in ability to convey understanding of the logic of the project, whereas critical path analysis is limited in other ways. We can compare a barchart with a PERTchart to evaluate their best roles.

Characteristic	Barchart	PERTchart
Timing	Shows timing clearly.	Does not show timing clearly.
Logic	Does not show logic.	Shows clearly which tasks depend on others.
Familiarity	Most people are used to barcharts.	Most people are not used to PERTcharts.
Presentation	Can be scaled down to A4 size.	Plans are hard to read and quite large.

There is a barchart on p. 49. Just spend a few minutes looking at it and you get a clear idea of what is going on in no time at all.

But, as my mother used to say as she put away her angling gear, 'Network analysis is a different kettle of fish.' Network analysis needs to be learned.

Network analysis – a primer

Network analysis is an entirely natural management tool. It is something that we all do every moment of our lives.

You normally open the kitchen door before stepping into the kitchen. You put on the kettle for tea before getting the cups out, as you know that boiling water is required for tea making.

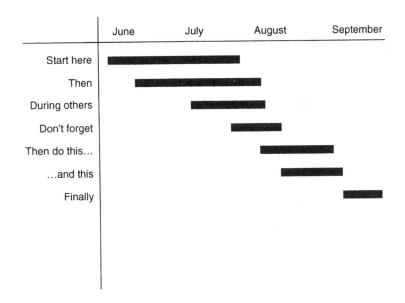

	June	July	August	September
Start here				
Then				
During others				
Don't forget				
Then do this...				
...and this				
Finally				

We unconsciously understand much about the flow of doing things, the logic that dictates the way things get done. Projects are just the same except that there are a larger number of tasks.

If you wanted to take your car engine out, you would read the book on the topic, and it would take you through various steps. It might explain which tasks had to happen before others and which could be done any time. You could probably figure out which bits your assistant could get on with under the car while you do some jobs topside. You can see that all the disconnecting work must be done before the hoisting can start.

What we can do is to break down any project into activities or tasks, and then decide how long each task will take and how each of these activities relate to one another. These relationships are the framework of the project. From these data we can calculate the timing of each element, and predict which activities are vital to the success of the project.

In general, the terms, **task** and **activity** are interchangeable.

The diagram on p. 50 shows the overall process.

You draw up a network diagram made up of tasks and links between tasks. At this stage you do not know when the tasks will occur; you know only how long each task will take and how each task depends on and controls other tasks. In fact you don't know how long the tasks will take; you have only got your best estimate – remember, we human beings are poor at predicting the future.

This network diagram can be analysed. This analysis is a process that we shall get to later, and it deduces from the model of the project when each task could and must start and finish. The timing of the activities is deduced from the network diagram. Additionally – here is a trailer for a forthcoming attraction – you can estimate how many resources each task will need. Based on these estimates, the process can add up the demand that the whole project will make for each resource for each day. But we are giving away the whole plot.

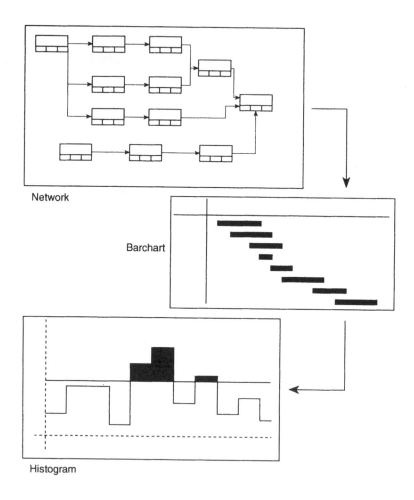

Network

Barchart

Histogram

All projects are made up of such frameworks. Because of the complexity of some projects, it is difficult to work out the timing calculations in our heads. We need a personal computer and a project-management software package.

There is no such thing as a project-management software package.

The generic term for these techniques is **critical path analysis** or **critical path method**, and there are two conventions that are commonly used. One is called **activity on arrow** (usually shortened to **arrow**) and the other is called **precedence**. Precedence is sometimes called **activity on node** and sometimes referred to as **PERT**. For the purposes of this book, these two techniques are called **activity on arrow** and **precedence**.

Both of these conventions aim to achieve the same objectives. They both expect a project manager or a project planner to break the project into **activities** and work out how the activities depend on each other. The process is then to calculate when each activity could, should, and must start, and to determine which activities are vital to the success of the project.

Some people become passionate when discussing the advantages and disadvantages of each convention. People go blue in the face defending activity on arrow, while others will stand by precedence until the cows come home. I don't mind very much. They are just like French and English, two different ways of saying the same thing.

The choice really is not very important. The chances are that you have had some project management software dumped in your lap and that software will support only one of the two conventions.

The vast majority of software systems are based on precedence.

The input to a critical path system is a list of activities such as 'weld pipes to tank' and the logic which defines what must happen before the pipes can be welded and what may happen after welding. The output is that welding the pipes could start on 17 August and must be finished by 27 August.

There are long, two- or three-day lectures on these topics, the sort of lectures where you can go, listen, try to stay awake, and emerge knowing less than you did before you went.

In fact, the subject is not hard; the concepts are as easy as not winning the pools.

The precedence convention

In the precedence convention you draw a box to represent an activity. The flow of time is generally considered to flow from left to right, but the size of the box bears no relationship to duration. It need not be a box – you can use triangles if you like, but boxes are the norm.

Therefore, if we had an activity called *dig trench* (activity number 16) with a duration of 5 days, we could draw it like this:

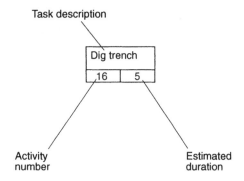

Task description

Dig trench

Activity number

Estimated duration

Let's draw two activities and connect them with a line. The line is called a **link, dependency,** or **relationship**, and it links activities together to show how they depend on each other.

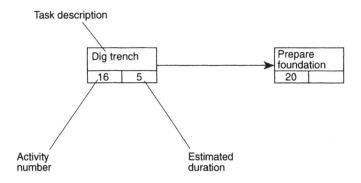

Task description

Dig trench

Prepare foundation

Activity number

Estimated duration

Wouldn't it be nice if everyone used the same terms?

Thus, we say that activity 16, *dig trench*, must precede activity 20, *prepare foundations*. Does this make sense?

The precedence convention dictates that the right-hand end of the box represents the finish of the activity, and the left-hand end represents the start of the activity. Therefore, in the two-activity example above, the relationship which connects the two activities says that we must finish activity 16 before we can start activity 20. This is a **finish-to-start** relationship, or a finish-to-start link, often called **F-S** by its friends.

Here is a very simple example. Let us imagine it is 07.30 on a weekday morning and you are about to set off to work. Your daughter is coincidentally about to leave for her sixth form college. The bike is locked away in the garage alongside the car and safe from your local branch of the Bicycle Liberation Front.

Once the garage door is opened you can cycle off to work whilst your daughter drives to college. Neither of you can leave until the door is open. It doesn't matter who opens the door but once it is open your departures are independent of each other.

In this case, *open the garage door* (task 6) precedes two other tasks. This means that both *get the car out* and *get the bike out* can start as soon as task 6 has been achieved. Therefore *get the car out* and *get the bike out* both succeed *open the garage door*. Neither of them can begin until *open the garage door* is completed.

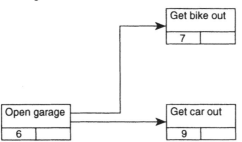

The succeeding tasks, which we can refer to as 7 and 9, are independent of each other. They can go on at the same time as each other, providing you have a car driver and a bike rider – either of whom could have opened the garage door.

This is fine as far as it goes, but not everything in the world is as simple as that. Sometimes we don't wait for one thing to finish before we start the next. Often we start one thing, wait a bit, and then start the next task, with the result that the two are running in parallel.

Take a look at this barchart showing the pipeline project in which the project management team are trying to lay a pipeline from the coast into a city. When the digging task has been going for a few days, we start the pipe-laying gang going. Then they are both working away. The trench diggers must stay in front of the pipe layers, who, in turn, must stay in front of the pipe welders, who start a little later.

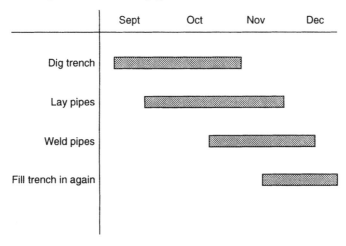

This is something that you may often wish to show. Wouldn't it be nice if there was a technique for showing it? Now look at the following little sketch – should that be the little following sketch? Either way, look at this:

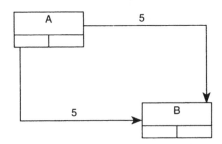

Start-to-start and finish-to-finish links.

We can draw relationships from the start of one activity (left side) to the start of another activity. And we can put durations on these relationships.

What the **start-to-start** (S-S) relationship shows is that 5 days after the start of the first activity we can start the second. We can also connect the finish of one task to the finish of another. The **finish-to-finish** (F-F) relationship says that when we have completed activity A, we expect to have 5 days left to do activity B. Useful. This is the real world. Consider the little network for a kerb-laying project shown below.

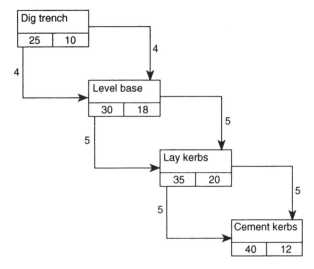

You can see that the overall duration of this little project is 26 days. The trench digging starts, and 4 days after that the base levelling gets going. After another 5 days the kerb-laying gang gets going, and 5 days behind them come the cementing team. They all work along together until each gang, in turn, finishes their part of the work.

There is a start-to-finish (S-F) link available in some conventions and with some software packages. I have never found a serious need for this but you might. The S-F link says that one task cannot finish until another task has started.

The use of S-S and F-F links, creating an overlapping of activities, is common and very difficult to show in the arrow convention. In the precedence convention, you may mix up start-to-start, finish-to-finish, the rare start-to-finish, and the very common finish-to-start relationships to build your plan.

The forward and backward passes

Once you have assembled a diagram showing the tasks in your plan and how they depend on each other, you can have the plan analysed. This is a pretty automatic process on most project planning tools, but it helps to understand what is going on under the bonnet.

The process begins with a forward pass. The first activity is given a starting date, and the computer adds the activity duration to the starting date to arrive at the earliest finish date for the activity – the earliest date when that activity could be completed. Each activity is taken in turn and its earliest start and finish dates are calculated.

The earliest start date of an activity is the day after the earliest finish of all preceding activities. The system looks at all preceding activities, finds which one finishes last and starts the succeeding activity on the next day.

If complex start-to-start and finish-to-finish links are used, the calculation gets a little more tricky. The analysis must calculate the start and end dates for each task via all possible routes – through the start links and through the task as well as through the end links. (See figure at top of p. 56.)

The program analyses all the activities in this way and eventually arrives at the end of the network, finding the earliest possible finish date of the last activity. Taking the earliest start of the first activity away from this date gives the shortest overall duration for the project. Starting at the end of the plan, the software now executes a backward pass, calculating the last date on which each activity can finish in time for all the succeeding activities. The program selects a task, finds its succeeding activities, finds which must start first, and gives the previous day as the latest finish of the activity in question.

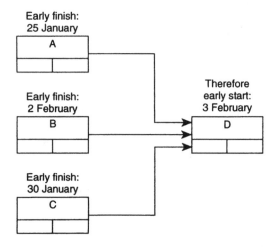

Once again the use of start-to-start links and finish-to-finish links complicates the analysis, making it necessary to calculate the routes through the tasks and through the links. Some activities will be found

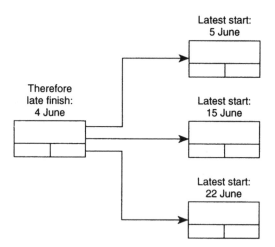

in which the latest and earliest finish dates are the same – these activities must happen on time if the project is to finish on time. Such activities are critical – they are critical to the timing of the project. Other tasks will have a window in time within which they should occur. The difference between latest finish time and earliest finish time equals the amount of float – the amount of time the task may be delayed before delaying the project.

It is possible in a precedence plan to have tasks with different floats on their start and finish dates. The start of a task might be critical, as getting that task started controls the start of a following task. Even though a task has no float on its start date, it can have float on its end date.

Imagine one fast task preceding a much slower task with S-S and F-F links. The fast one might have a critical start, as it controls the start of the slow one. In reality the fast task rushes ahead of the slow one, with the result that the critical path passes through the second task. This gives the first task some float on its end date. The end of the first task can be delayed, without affecting the end of the project.

Thus, in the precedence method, we have to be prepared to deal with start floats and end floats and to recognize that these may be different. You can only get into this arena by using start-to-start and finish-to-finish links. A plan made up of simple finish to start links avoids this complexity altogether.

There are two types of float. While we are thinking deeply about such things, we had better deal with the idea of **free float** and **total float**. Free float is the amount of time a task can be delayed without affecting any other task, whereas total float is the amount of time a task can be delayed without affecting the end date of the project.

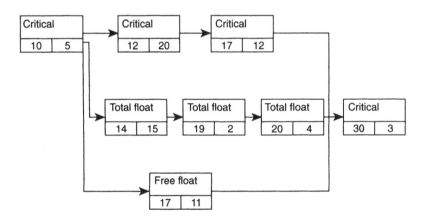

In the diagram, there is a chain of tasks which share float between them. Task 14 can be delayed, but, if it is delayed, it eats up the float of tasks 19 and 20. So task 14 has total float but not free float. Task 17, on the other hand, can be delayed for 10 days without any adverse effects; therefore, it has both free and total float.

As soon as this analysis process is completed, a barchart can be printed showing the earliest start and the earliest and latest finish. This can be compared with the original objectives and any problems can be evaluated.

That's all, folks. That was critical path analysis. In a few short pages you have gone from the basic components of network diagrams to some of the most erudite topics. This is all rather theoretical, and we cannot let our invisible project manager forget reality; therefore in the next chapter we shall try to demonstrate the techniques in reality. We had better take a look at choosing tasks first, and then we shall sit in as a group of planners get down to business and create a plan of their own.

Choosing the tasks

What are tasks like? One of the problems that many project managers have is breaking down the project into its activities. Clearly, one task can represent a project like 'Build Bridge'. An activity can be minute, for example, 'pick up screwdriver'.

Tasks are your choice. They are the steps along the path to the end of the project. They are normally defined by three bits of information – the task identifier, the task description, and the task duration.

The task identifier is really only between you and your software package. It's a handy reference that you might use from time to time to save typing out the description every five minutes. Some software systems hand out these identifiers for you; some allow you to type them in. Some systems allow only numbers; some allow letters and numbers.

Tasks can also have dates and other bits of information associated with them such as organizational codes, responsibility codes, resources, and costs, of which more later.

Tasks can be about anything. Long tasks include building a new bridge with a duration of 10 years. Short tasks include dipping a printed circuit board into acid for 10 microseconds. Tasks always take time – if they take no time there is little point in worrying about them. Tasks may not involve much activity. Paint-drying and grass-growing are very reasonable and normal tasks. A very common task is waiting for delivery. Particularly in the precedence technique, it is a common error to put these sorts of not-doing-much items in as links.

Look at the figure on p. 59. Here the planner has a five-minute task called *order pizza*, and then a 20-minute delay before a second task called *eat pizza*. The barchart, which is the primary communication document, shows only the two tasks separated by a 20-minute gap. The first question will be – why the gap? It is much better to show an *await delivery* task so that everyone knows what is going on.

There are some guidelines to help you select useful tasks but there are no definite rules. Remember that you can always go back to your project plan and add more tasks to examine some aspects in greater detail – you need not spend too much time trying to get it right first time.

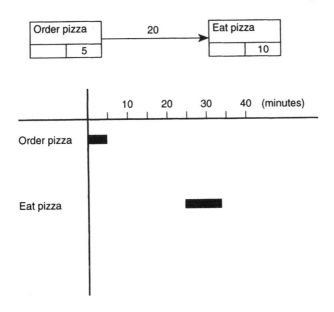

Let the detail grow as you learn about the project. When project planners used mainframe computers, you could only get on to them once in a blue moon. Hence you had to make a detailed plan from the word go. Now you have a personal computer sitting on your desk, and you can build the plan as you feel ready to do so.

If you don't know much about the telephone system that is to be installed in your new offices, add a task called *install phone system* for the moment. Later on, when the detail emerges, you can add extra tasks to look more closely at this stage.

Normally plans have a fair amount of detail showing the start of the project and the early work that has to be done. Often the future is hazy and is summed up in a few broad-brush tasks.

Experience is the facility we use to select our tasks, but here are a few short rules about choosing tasks.

Include the whole project – it would be beneficial if you break down the whole project into tasks, missing out as little as possible. Start with the overall task – Launch the new light-bulb – and then start to break it down into tasks.

Choose tasks that you can define – if you cannot describe it reasonably easily, then you probably have not got a very good task.

Try breaking it down into two or more tasks. Try joining it to another element of the project. It is useful to define the activity in order that your colleagues can understand which parts of the project this specific activity represents.

Choose tasks of similar time scales – roughly speaking, the more tasks you have, the shorter their duration will be. In a 10-year project represented by two tasks, the average duration will be at least 5 years. In a 10-year project with 1000 tasks, the average duration will be around 3 or 4 weeks. Very long tasks do not work well alongside very short ones. Imagine a 1-day activity under another that takes 1 year.

Activity durations should relate to the update cycle – if you are updating weekly, then very little will change on a task taking 3 months. A very rough rule of thumb is that activities should be from 2 to 5% of the project's duration. On normal projects most of your activities will be between 3 days and 3 weeks. Longer activities may be used to indicate work executed by others.

The description that you choose can make a huge difference – it is the prime chance to communicate what the task involves. This is the piece of data that will almost certainly appear on your reports. Descriptions can be coded; indeed, this is traditional in the project planning industry. Descriptions often sound like this:

Bld SW. Wall grid 34–56 inc plnths

This is very useful if you want your project to be like an adventure game in which everyone is kept in the dark, but if you want to communicate with your peers, clear descriptions will help.

You can use the activity description to explain as much as possible about the work and add additional notes if you think it necessary. Much confusion occurs when the activity description is not clear. Misunderstanding of activity content causes problems in assessing durations, and progress monitoring becomes confused. Let's take an example. *Lay road surface* might or might not include road-line painting, kerbing, and, for that matter, tarmacking.

A classic example is *order specialist item*. Does this mean actually writing an order? Or does it include all the preceding steps such as getting quotes and choosing the supplier? Does it include the succeeding items, such as awaiting delivery, monitoring manufacture, and the transportation of the item to your project?

It is a little risky to assume that the content of a task will be made clear by the other activities around it. It may be that you will produce reports that do not include those other activities.

Some suitable descriptions include the following:

- build south wall;
- weld platform to leg #3;
- construct brickwork to south wall;
- code input routine;
- install window frames to first floor;

- design dashboard layout;
- delay time waiting for advertisements to appear.

We have already had *order pizza* so we should have *digest pizza, clear up floor*, and *visit doctor*.

Note that these task descriptions have a verb in them, what school-teachers call 'doing' words.

The loop

In project planning you may meet a problem for which you should be prepared. It is called a loop. In a loop a path through the network diagram loops back on to itself and therefore creates a plan that cannot be analysed. The analysis routine goes round and round until it gets tired and gives up. Here is an example of a loop.

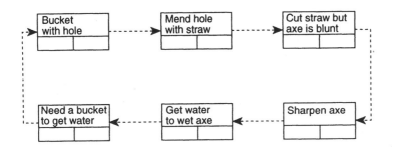

Drawing up the plan

In order to ensure that you do not become too invisible and to help you in an extremely practical way, here are some tips for drawing up the network plan itself.

Firstly, is it a good idea to draw up the plan? There are those who argue that drawing up a plan is no longer necessary. They argue that since you will be building your model within the memory of the computer, what is the point of drawing the whole thing up beautifully on paper? What such people overlook is that 50% of project planning is about thinking ahead, and drawing up the plan is a useful part of that process.

Getting the key members of the team around a table, a wall, or a whiteboard can be very productive indeed. Much discussion ensues about the how, the what with, and the when of the project. Decisions are made; questions form. It is an unstructured opportunity to think about the project.

That doesn't mean you are going to spend hours and hours poring over a hot drawing-board and producing the world's neatest critical path diagram. Your software package will do that for you. All you need to do is arrive at a list of tasks and links in order that you can type them into your software. You can do this planning stage more than once. You might subdivide the project into a few natural stages and get the right team together to draw up a network for each stage. Throwing all the plans at the computer and putting in the links between the various groups of tasks gives you a master plan that everyone will be happy with.

This is the way to make an impression early on in the project's life. We can conclude that you will be drawing up plans, probably in the company of others. Here are some tips for doing that.

There are discrete steps in drawing up a PERTchart. (I know – I keep throwing in different terms, but I am trying to make sure that you know them all.)

Here is a network diagram for drawing up a network diagram.

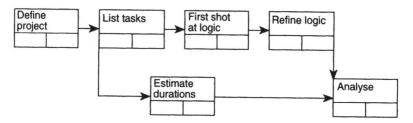

Network for drawing networks.

You first define the project and then get a rough list of the tasks – this is usually the result of a brainstorming session. The tasks can be in any order at this stage. Next you have a first shot at drawing the tasks in logical sequence. This usually results in more tasks being identified and loads of discussion about the how, the when, and the where. If the project is already in trouble it may result in white faces scuttling for cover as they realize what they have let themselves in for. At least they are learning and talking about the project – you've got them thinking ahead. The resulting logic diagram (or PERTchart or network diagram) normally looks like a spider fell into the inkwell and then walked all over the diagram in a drunken stupor.

Meanwhile someone might have been calculating how long these activities are likely to take – estimating durations it's called. Then you are ready to tap the whole lot into the computer and get the plan analysed. A quick inspection of the resulting barchart may have you scuttling off, looking for a new job before everyone else realizes how bad things are.

How about the actual drawing stage? Some people use a

whiteboard. They gather around the whiteboard and start drawing tasks and links with coloured markers. This involves plentiful use of the wiper, and descriptions tend to get abbreviated, as those broadtipped pens are not ideal for small lettering.

An amateur survey indicates that people standing up looking at a vertical surface draw better plans than those sitting down looking at a horizontal one. This may have something to do with perspective. Of four people sitting around a table, two see the plan the right way up and two see it upside down.

Some people get a large sheet of paper – the back of an old plan or architectural drawing – and start on that with a pencil and a rubber. Some organizations have paper printed up for drawing plans on. The sheet of paper is printed with a grid of, say, 100 task boxes (10 by 10), so that all the team have to do is write the descriptions into the boxes and draw the links between them. Some people have a rubber stamp made up with the task box raised on its face in order that they can quickly, if rather irreversibly, stamp tasks all over the place.

A very popular technique is the Post-it technique. You write the tasks on Post-its, one task per sheet, and stick them on the whiteboard, table, or other smooth surface. By placing them in neat lines and next to each other, you can portray most of the logic. You can draw the extra lines on with a pen or pencil.

This works very well and will survive the minor gale caused by the engineering director's sudden entrance. Bits of paper or card are easily disturbed by such gusts destroying your work and personal image in a puff. When the plan makes sense, you can write the succeeders alongside the description on each Post-it, stack 'em up, and type them into the computer. Neat or what?

Time to relax, and grow wings and a proboscis. In the next chapter you can be a fly on the wall while a small group of people try to draw up a network for a real project.

A fly on the wall

In this chapter the reader can adopt the unlikely role of a fly on the wall and listen to a small group of people assembling a plan for a project. The project management team is fictional and bears no resemblance to persons living or dead, and the project is as realistic as it is hypothetical.

Fly on the Wall. You are that fly. You are currently hanging upside down on the ceiling in the conference room of Squirtem PLC – the renowned manufacturer of pumps.

The managing director (PJ) has decided that the company will be represented this year at PumpEx International – the International Exhibition of Pumps, which this year is to be held in The Netherlands at the Rotterdam Exhibition Centre. Squirtem PLC has exhibited at shows before but never overseas. This year the board has decided to launch the new Squirtem 4000 pump at the show and make a big splash about it. This year PJ has asked that everyone try to avoid the mad panic that normally happens just before a show.

Pamela Project has been selected to manage the whole project and she has got the engineering director (Stanley Knife) and the export director (Ove Seas) together to discuss the project, well in advance of the exhibition. The date of the meeting is 5 January 2006 – they have just returned from their Christmas holidays.

Pamela's prime objectives at this meeting are to determine what has to be done, to get the team thinking ahead about the project, and to produce a network diagram for it. Your objective is to land on that delicious-looking ham sandwich that was dropped on the floor at the end of the conference table at lunch-time.

Draw your plans on flypaper.

You drag your attention away from the ham sandwich and eavesdrop on the three people sitting around the conference room. This is what you hear.

Pamela is just finishing a five-minute overview of project management.

Pamela: So that about sums up project planning – do you think we can try to do a network plan for the show?

Stan: We'll give it a go – it sounds useful.

Ove: Ya – ze last time ve vent to a show it was a near disaster – anything zat can help has just got to be a gut think. Vere do ve start?

Pamela: First we must define the project.

Stan: That's easy – we want to exhibit the new Squirtem 4000 pump at PumpEx in Rotterdam.

Pamela: That's good – but do we have a pump, do we have a stand, how are we to get the pump there, are we doing some leaflets or handouts, has anyone booked the space at the show and the hotels? I know that we have a time limit – the date of the show – but do we have a cost target as well?

Ove: Ah, zose are gut questions. I have a budget for transporting the pump, the stand space, and the hotels, so I don't zink those have been done yet. Ve definitely want a brochure, and it should be in about four languages. How about a pump, Stan?

Stan: I have the prototype in bits in the design studio at the moment and about four preproduction models on the shop floor – when do we need them by?

Pamela: We don't know yet but first let's look at the processes – can we get a first breakdown – at least a list of headings?

Ove: OK, OK, Ya. I can see some headings, they are: brochure, accommodation, stand, transportation, and er –

Stan: Prototype?

Pamela: OK, let's take them one at a time – let's try brochure.

Ove: Ve need to get the text written and then translated. Ve need photographs of the pump, and then ve need to find a printing company, and ve need to establish a number of copies. Finally ve need to get the printing run off before the show.

Pamela: Surely we need to have the brochures printed some time before the show to allow for shipping them over to Holland.

Ove: Ve could get the stuff translated and printed in Holland – zey have good translation companies there.

Stan: That's a good idea – we should get guide prices from both the UK and Holland to check the cost, but we could do that right now, couldn't we?

Ove: Ya, ve could. Ve would need a specification first so that the prices are for the same zing.

Pamela: Hold on a second – let's back up a moment – here is my task list for the brochure part so far.

Moving across the ceiling, edging a little closer to the sandwich, you find yourself directly over Pamela's list – this is what it says.

Brochure

- prepare specification for prices;
- get printing prices from UK and Holland;
- get translation prices from UK and Holland;
- check delivery costs;
- check customs requirements;
- write text;
- photograph prototype;
- prepare artwork;
- print brochure;
- select printing company.

Stan: Hey, we need to have a prototype before we can take pictures of it.

Pamela: That's a good one – I bet that is on the critical path – how about a nice drawing done by an artist, could that be done anytime?

Ove: Let's keep that one up our cuff in case we need it – I would like a photo in the brochure if possible.

Stan: We say 'up our sleeve', Ove, not 'up our cuff'. I hope you don't mind my correcting your English.

Ove: Sank you. It is much better I learn than make the same mistakes time and over. You have just done a list of tasks – you have not said how some tasks depend on others?

Pamela: No, not yet. Let's concentrate on getting a list of tasks first and then we shall turn to the logic. We'll assume that we take a photograph of the prototype as soon as it's ready, for the moment. If that causes a problem we have the drawing as something to fall back on. We should know if there is a problem as soon as I get this lot through the software. How about the prototype, Stan?

Stan: We have to reassemble it. We took it to pieces to check the bypass valve assembly, which turned out to be fine. I would like to paint it ready for the show. So my stuff is pretty linear. Paint the components, reassemble, pack it away in a crate, and ship it to Holland in time for the show.

Pamela: Do we need customs paperwork for the pump?

Stan: Oh yes, I forgot that. We need to apply for a *carnet de passage*, which lists all the bits that are being sent. That takes about two weeks to get hold of.

Pamela: Anything stopping us doing that right now?

Stan: No. . . . Oh yes, we need to book a shipping agent first.

Ove: I vould like to take a picture of the new pump leaving the factory so zat we can use that in our publicity material. I bet our beloved managing director vill vant to be in ze photograph when the pump goes out. Hey, talking about PJ, vill he vant to approve ze brochure?

Pamela: I bet he will. So I'll add 'approve brochure' to the brochure list, and here is my list of stuff for the pump.

Pump

- paint prototype components;
- reassemble prototype;
- appoint shipping agent;
- get shipping paperwork;
- get packaging materials;
- pack up pump;
- photograph pump leaving the workshop;
- ship pump.

Ove: Great – ve get a photograph of the pump in a packing case – ve vill have to mock up the photograph before the pump gets packed away. Has anyone got any flyspray? Zat insect is really annoying.

Pamela: Let's move on to accommodation.

Stan: This really helps, Pam, we are solving lots of problems long before they come up, I'm really glad we had the chance for this chat.

Pamela: Thanks, Stan. As far as accommodation goes, I think it breaks down into two categories – the stand and hotels.

Ove: Yes, ve need a stand designer first, and when ve have got one ve need the design for the stand itself. I have a budget for this, but ve need to write a brief telling the designer vot ve vant to do. Ve also need to book space at the show. It says in the show's preliminary information pack zat ze last date for space booking is 27 January.

Stan: Don't we need to decide how much space we want before we book it?

Pamela: That's a circle – we need to know how much space we want before we can book it and for that matter design it, and we need a design before we know how much space we want!

Ove: There are only two choices – ze 10-metre or ze 20-metre stand. Ve have enough in the budget for a large stand.

Pamela: What about building the stand itself?

Ove: Ya, ve need to select and employ a stand construction company in Holland – zat is something I can do. They will take care of all ze stand, the carpets, the services, and hire the, er, vot do you say – chairs and tables and zo on?

Pamela: Furniture?

Ove: Ya. Of course. I will need to have the design before I can get prices.

Pamela: We also need to finalize a list of how many people are going so that we can book hotels and plane tickets. So here is the accommodation task list.

Through your many-faceted eyes you see the list that Pam has produced.

Stand

- select stand designer;
- prepare design brief;
- design stand;
- approve stand;
- invite tenders for stand construction;
- place order for stand construction;
- build stand;

- book space (before 27 January);
- finalize list of staff;
- book airlines;
- book hotels.

Ove: Zat is the first time you have mentioned real dates.

Pamela: Yes, that's the first externally controlled date – whatever happens we must hit that date.

Ove: In that case don't forget zat ve cannot start building the stand until the veek before the show – 9 April.

Pamela adds '(Not before 9 April)' alongside the build-stand task.

Pamela: OK, let's take a five-minute break and then get down to logic.

Five minutes later. You have managed a quick nibble of the sandwich while everyone was out of the room, but you have become a bit concerned by the quantity of sticky yellow bits of paper that Pamela has brought in with her this time. These look alarmingly like flypaper.

Pamela: OK, I have written each task on to these Post-it stickers. Let's try and assemble the network diagram, taking the tasks in sequence. Let's start with the brochure tasks. The way I see it, we could *prepare a specification for printing prices, check delivery costs, check customs requirements*, and *start to write text* immediately if we wanted to. So all those tasks can become **start** tasks.

Ove: Vot does that mean?

Pamela: It just means that these tasks do not depend on any others. Now when we have the printing specification we can *get printing prices* from the UK and Holland and at the same time *get translation prices* from the UK and Holland.

Stan: Yes, that's right. They both depend on the specification, but they are independent of each other. We could prepare the artwork, leaving space for the photo as soon as we have the text translated – hang on, we didn't have a task called *translate text* – can you do one please, Pam. Also it would be better to know the specification of the brochure before we start writing and talking to customs.

Pamela prepares the missing Post-it.

Stan: OK, we can translate the text once we have the prices back. No, wait, we must select the translation company *(Pam prepares another yellow slip)* after we have the prices, and then after the translation is finished we can print the brochure.

Pamela: Surely we need PJ's approval before we can print and we need to *photograph prototype* before we can *print brochure*. The task *select printing company* is left over. We can select the printer once we have the prices, and we must have a printer before we can print the brochure. Let's take a look at that so far.

They all stare at the whiteboard. The whiteboard stares back.

Stan: The task *print brochure* doesn't link to anything at the moment.

Pamela: You're right. What depends on that? What tasks cannot happen until the brochure is printed?

Ove: The show – ve need the brochure at least one veek before ze show so that ve have a little time up our sleeve in case ze printer is late.

Pamela: That's fine.

You flit across to the opposite wall and take a peek at the network diagram that is beginning to emerge on the conference-room whiteboard. You begin to think that the afternoon tea might arrive soon and that usually means a bowl of lovely sugar. This is remarkably good forward planning for a fly.

(See figure on p. 71.)

They carry on working their way through the other categories until the network is nearly ready.

Pamela: Now we need to think about timing. How long will it take to get a specification for the brochure prepared?

Ove: I guess zat I should know zat one – the last time I did one it only took a day to do but zen it had to be typed up and sent around – let's say one veek.

Pamela: How about *getting quotes*?

Ove: In Britain zat takes about two veeks; in Holland ve expect a price in four days.

Stan: Surely it makes sense to allow two weeks as the worst case.

Pamela: Absolutely – we'll see if that causes a problem – but I don't expect it will be critical.

Stan: Are we talking about real days or working days here?

Pamela: We are talking working days. I will tell the program that we work five days a week and that we take a week off at Christmas and a day or so near Easter. It will sort out when things should happen from our durations and logic.

They press on until they have arrived at a complete network of tasks each of which has a duration and a task description.

(See figure on p. 72.)

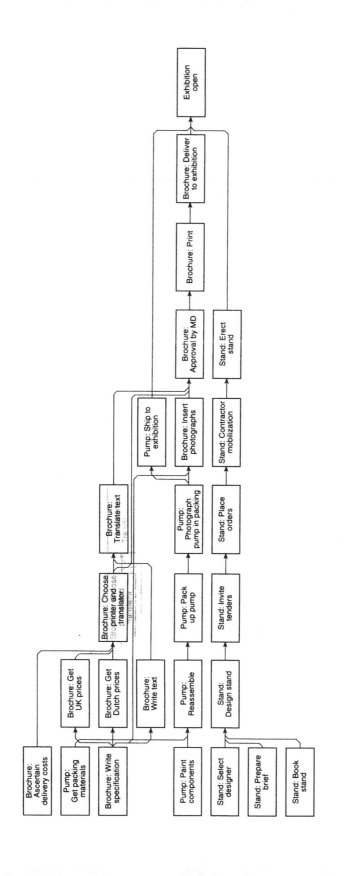

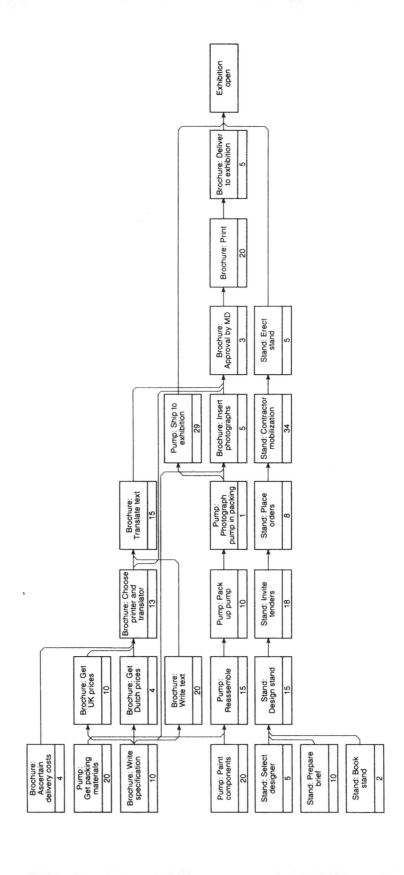

Pamela: We are ready to type this into the computer. Do you want to watch, or shall we get together tomorrow to see the results?

Ove: I vould like to vatch, please.

Stan: I'll leave it to you two. Give me a call when you are through.

They type the plan into a software package and print out the first barchart, and this is what they got:

(See figure on p. 74.)

Pamela: There seems to be a slight problem – we are going to be all alone in the exhibition about three weeks after everyone else has gone home. We'd better get Stan back.

Pamela phones Stan while Ove looks at the barchart.

Ove: I do not understand ze problem.

Pamela: If we do the project just as we said, we shall be three weeks late for the show. Look at the critical path. Firstly, we won't be ready with the pump, and, secondly, the brochure will be late.

Ove: It is time to get some of zose things out of ze sleeve.

Stan arrives and takes a look at the barchart.

Stan (thoughtfully): You know it's a great deal better to find this out in January than in April!

Pamela: That's true. Anyway, if we do the translation and printing in Holland, we can reduce the duration for delivery to, say, two days and reduce the duration of the quotation down to four.

Stan: This may not be the cheapest way.

Pamela: That's right, we might be taking an expensive route so as to get there on time.

Ove: It should not cost much more.

Pamela types in the changes.

(See figure on p. 75.)

Stan: That's no good. Does this thing change, I mean recalculate the whole plan when you make a change? How about the prototype? You have changed the durations but the stand is still not erected in time according to the new barchart.

Pamela: Your area is OK, Stan – it's the stand that is causing the problem now. You see how the critical path has moved? We must cut some time out of the stand contractor's schedule to bring the date back.

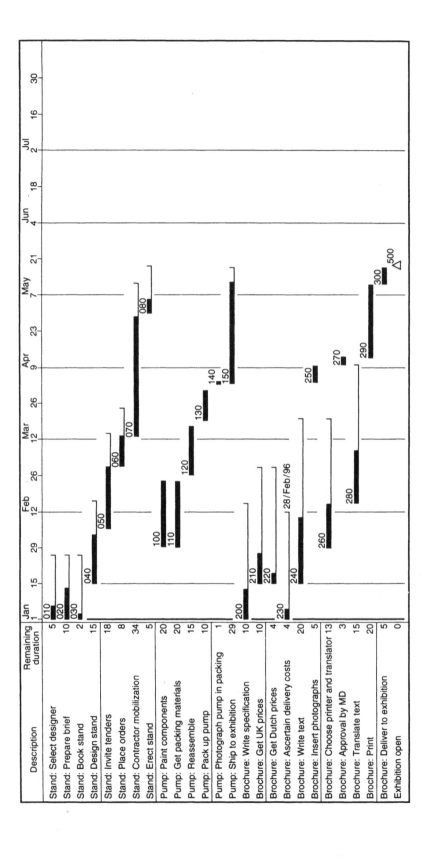

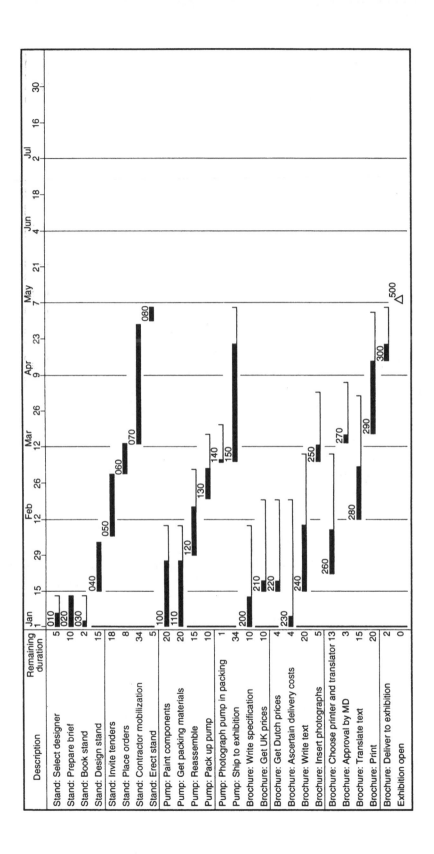

Ove: I zee. When you changed the schedule for the brochure, the stand erection became the most important, the most critical area?

Pamela: Yes, that's right. Can we take two weeks off the stand contractor's mobilization time?

Ove: Ve can if ve say so in the initial invitation documents – ve couldn't if ve tried to bring in a change like zat. Will zat fix it up?

Pamela: I think so – here is the revised barchart.

Pamela: That's fine, isn't it? As long as PJ manages to approve the design in three days we should be OK. We'll go with this plan?

(See figure on p. 77.)

Stan: Suits me.

Ove: Ya, can I have a copy of ze barchart to keep with me?

Pamela: Of course you can. I shall print a copy of the barchart for each of us. Come to think of it, I'll send a copy to PJ for his information, and I'll tell him what he has to do and when. I have marked each task with a code name like *brochure*, or *pump*, or whatever so that I can print off a barchart for one group of tasks if you want.

Ove aims a swat at a passing fly and misses.

Stan: I don't think that is necessary – it is pretty small – although I can see the benefit of that if the plan gets to be very big. Even in a small project we have to know who is doing what. I am tired of projects where half the jobs get done twice and the other half never get done at all. Are we clear who is doing what on this project?

Pamela: I think so. I have put our initials against each task to show who is the prime mover for each task – take a look at this display and see what you think.

They look and decide that Pamela has got it right.

Stan: But, surely, things will not go exactly like this. What do we do when things change?

Ove: Now zat's a gut question.

Pamela: From time to time, we should get together, review progress, get the latest news, and update the plan. Updating means taking each task in turn and modifying it to represent the latest situation. Then we have the plan analysed and decide what we should be doing. Finally, we reprint the barcharts and send them to everyone interested.

Ove: Vot about money?

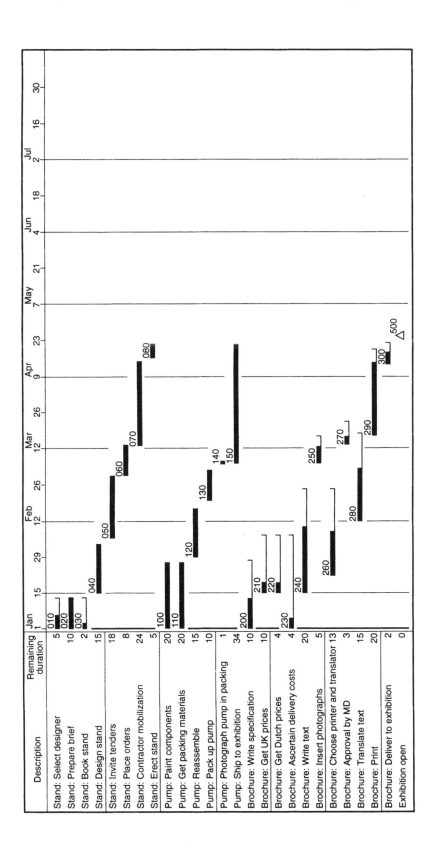

Pamela: Yes. We could estimate the cost of each task, enter the cost estimate, and get a cash flow curve telling us what we need to spend and when. This will give us a budget as well.

Using the recently printed barchart, Ove swats the fly that has been annoying him throughout the meeting, and so our story endsssssssssssssssss.

Resource management

Within this chapter the humble pages offer written explanations of the techniques used to plan and reconcile resource problems on a project. Histograms are drawn and analysed, and resource types and their methods of analysis are also examined.

Resources are what go into producing work. Things like bricklayers, welders, carpenters, cranes, Sue, foreman tree fellers, and authors are resources that produce work. Things like bricks, tiles, widgets (they get everywhere, don't they?), computers, concrete and rivets are resources that go into work.

> You are a resource.

In the world of project management all such things are resources.

This discussion assumes that you have already got a model of your project, perhaps in a network diagram, certainly within a computer. The sorts of things we shall be discussing are impractical without a computer to help with the simple but voluminous mathematics.

What are project planners trying to achieve when they start playing the resource game? Firstly, you can predict the demand the project will make on your resources. You can estimate how many carpenters, computers, or cows you will need. You can predict bottlenecks – you can foresee a time when the project will demand more resources than you can get. At such a time the project's progress will be held up by the lack of a precious resource. It is exactly the sort of problem that is worth while seeing in advance so that something can be done about it.

You can do your best to make sure that the way you proceed is such that the project makes the minimum demand on the resources and that the resources are kept busy. In project management terms this is called resource smoothing – trying to ensure that there is a reasonably smooth demand for major resources, not a series of sudden high demands followed by very low demands. Resource demands that are not smooth mean that you will need a large number of resources many of whom will be sitting around one day but working really hard the next.

Finally, you can experiment with different ways of proceeding with the project. Using loads of resources gives you a fast but expensive project; using very few may give you a slow but cheap project. There is a huge range of answers between these extremes and you can estimate some of these. Based on your calculations, you, along with your management, can make rational decisions about the right way of attack for your project.

> Resource planning is mostly about foreseeing problems.

There are five major stages when dealing with resources:

- resource definition;
- resource allocation;
- resource aggregation;
- resource aggravation;
- resource levelling;
- resource smoothing.

Who spotted that there are six stages listed? Resource aggravation is not a real stage, but the aggregation of resources can be very aggravating indeed. Let's take a look at these topics.

Resource definition

This is the first stage, in which you decide what resources you are going to deal with. At first it is very tempting to try to look after all the different resources that you can think of. Later on, you will select a smaller group of important resources – the ones that you can track and that are worthwhile tracking. It reduces the amount of work considerably.

People are very ingenious in finding resources to track. Much lateral thinking goes on in some project managers' minds in attempts to find resources to track that will benefit the project. Here are just a few ideas.

The obvious resources are worker types, correctly known as skill types: carpenters, computer programmers, welders. Here we say that a task needs, for example, four welders for each day in the task, or that we shall allocate two engineers to that task. This means two engineers for the period of time that the task will take. In such an arrangement the tasks are driven or controlled by the duration, and the demand for the resources follows on from that.

The second type are worker-time-units: carpenter-days, computer-programmer-weeks, engineer-days, machine-hours. Here we are dealing with the effort that a skill type can put into the task over a given period of time. If we say a task needs 120 machine-hours, we are estimating the total amount of effort needed to achieve that task. It is not 120 machine-hours per day – it is 120 machine-hours in the task in total. One machine takes 120 hours, 4 machines take 30 hours.

> One hundred and twenty machines might not do it in 1 hour.

In these cases we have estimated the work content, and the task duration is derived from the work content and the number of resources available. Some software packages support exactly this sort of thing and allow you to use one or both of these two major types of resource definitions. Some software systems make you choose when you introduce the resource; some allow you to allocate the resources in both ways.

Resources can be regarded as consumable and non-consumable. The difference is fundamental. A consumable resource is consumed as it goes into a task. Money, concrete, engine-days, and paint are all consumable. Once a banknote is spent or a gallon of paint splashed on the wall, that money or paint has been consumed. You cannot use it on another task.

> There are plenty of project managers who believe that nine women can have a baby in one month.

Non-consumable resources can be used, and used, and used again. You are like this. Once you have done 4 days' work on a task you can carry on work on something else. So can a crane, a painter, and a computer. The difference between these two types of resource will become obvious when we come to reports.

There are other bright ideas that project managers use. In aeroplane maintenance, space is at a premium. When a jumbo jet is serviced, there can only ever be three workers on the flight deck just because the place is so small. The flight deck, by the way, is where the driver and his friends sit when you are flying off to the Costa Plenti. To allow for this, the ingenious planners in aeroplane servicing use a resource called 'one-third of a flight deck'. Of course there can never be more than three of these in use at any one time.

On North Sea oil rigs, there are many types of workers. There are electricians, drillers, catering staff, and cleaners. There is probably a whole host of other specialized workers whose role we would not understand at all. They all share two things in common. They cannot go home for the night, and they all like a bed to sleep in. No matter how high-tech their function, when it comes to bed-i-byes, they are all the same. Hence planners on oil-rig projects use a resource called bed-space. This resource is strictly limited by the number of available beds.

Incidentally, a clever Swede, recognizing bedspace as a limiting factor, designed a bed that turns over, spinning on its axis like a spit. This 'double bed' could be used twice a day (by two different shifts) without the need for fresh sheets every 12 hours.

Is time a resource?

Often a resource definition includes an estimate of how many of each resource will be available to perform work on your project. This is called **resource availability**. Resource availability may be a simple number – 9 engineers would be a simple example. This can be more complex, as some project management systems allow the creation of resource availability profiles.

Resource availability profiles are graphs that show how much of a specific resource will be available over a period of time. They show details like 10 welders in June, 15 in July, and 12 until 17 August. When you create resource profiles, you need not worry about holidays and weekends, as the profile will be overruled by the calendars, which show when people work and do not work.

Resource allocation

Next comes the stage where each resource is allocated to some of the tasks in your plan. On the basis of the duration and the work content, you take each task in turn and decide how many resources that task needs. A task can have many resources allocated to it; for example, a task described as *weld pipes to calorifier* might require 3 welders, 2 welder assistants, 1 welding rig, and 200 lengths of pipe.

Each task is taken in complete isolation – you do not know whether you are giving the task too many resources and doing it unnecessarily quickly, or whether speeding it up would make the project as a whole noticeably faster. At this stage you just don't know.

You may allocate resources to tasks in many ingenious ways – if your software permits. Some of the more complex systems allow you to say that a task needs 4 welders for the first week and then 2 welders for the second, plus a crane for the first 3 days. Such a varying demand for resources over the length of a task is called a **resource demand profile**.

However you decide to allocate resources to tasks, you will be saying that you believe that the particular task will need so many tradespeople and so much material to get it finished. You will probably find it convenient to track only the key resources. It is not too hard to predict the demand for resources on a task-by-task basis. What is hard, without a computer, is to predict the demand for resources over a whole project. That comes next.

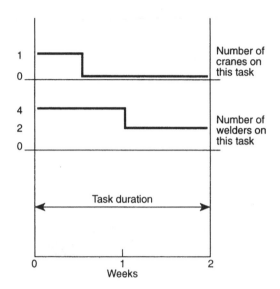

Resource demand profile.

Resource aggregation

In the twinkling of an eye the software will total the demand for each resource on a day-by-day basis and report, in a histogram, this demand. You can then peek into the future and compare the estimated requirement for your resources with their availability.

All that happens is that the software refers to the barchart, which tells it when the tasks are planned to be going on, and the resource allocation, which gives it a demand for each resource for each task. Taking each day in turn, the software adds up the demand for each resource. If resources should be beavering away on a task on that day, its resources get added in to the total. You could do this by hand but it would take much longer.

Resource aggravation

If you need 26 welders in December and you have 30 – fine. If you need 19 programmers in July and you have only 10 – problem. There is not really a stage called resource aggravation, but you get the message

– once again you have become visible and are able to predict problems. You can see the problems while they are small and distant and very easy to resolve. You may easily find the need to discuss this with the powers that be, achieving two things by the discussion:

1. resolving or at least recognizing the problems;
2. improving your reputation within the company.

The computer cannot solve these problems for you, but it will let you know about the problems ahead of time – in time to do something about them.

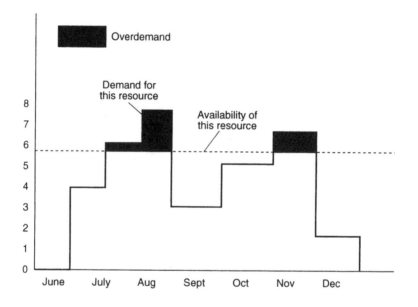

Resource levelling

Once again we have encountered some rather ill-defined terms. The terms used here may not coincide with the ones used by your software package or your fellow project managers. Yes, there is a British Standard of project management terms, but it is not very helpful.

Resource levelling is also known as resource limiting or resource limited scheduling. Whatever its name is, **resource levelling** is a process by which the software ensures that the project never demands more resources than you have said are available. After levelling, resource demand never exceeds resource availability. When you ask the software to level your resources, it will use the resource availability levels as a target to aim for and try to find a way of achieving the project without exceeding the resource profiles. Tasks are delayed automatically so as to wait for resources to become available.

This is what happens. Normally, in a simple time analysis, the software just checks to see that all preceding tasks are complete before beginning the next. In the case of resources levelling, the software checks to make sure that all preceding tasks are complete and that enough resources are available to do the work. If the resources are not available, the system delays the task until the required resources are available.

Two thoughts spring from this into your ever alert mind. Wake up at the back there. Firstly, delaying a task until resources are available may well cause other tasks to become delayed well past their earliest possible starting dates. If a task is delayed until resources become available, other tasks that depend on it must also be delayed. The end result may be a much prolonged project, In other words, the price of a strict adherence to set resource availability limits may well be a longer project.

Secondly, as the computer program takes the tasks in turn and tries to position them in time, the ones that get picked first on a specific date have a much better chance of getting the resources they need. Imagine a number of tasks, all of which use the same resources, all happening on the same week. Some start in that week, some end in that week, and some pass right through the week. Which tasks get the precious resources first, thereby leaving the leftovers for the other tasks?

Tasks taken later on have only the resources left over from the ones taken earlier. It follows that the sequence of choosing the tasks is very important. Most systems deal with critical path tasks first. Some project management software packages allow you to control what is going on by a prioritization system.

You may recall that a few pages ago I said that you could expect the software to calculate the resource demand for a task based on work content. Some packages allow you to enter the work content (6 programmer-weeks) against each task, and then the program allocates programmers to the tasks arriving at a duration for each task and hence a duration for the project. You can see that this is more powerful than defining a duration for each task yourself and having the software work out the resource demand from your duration and your estimate of work content.

If you use such a feature – sometimes called a duration scheduler – this prioritization problem is even larger. Additionally, you will normally be able to enter practical limits for each duration beyond which the system will not pass. A task that involves erecting a circus tent might involve 20 rigger-days and would be stupid if fewer than 4 or more than 20 riggers were working on it. In that case the acceptable duration range would be 1 to 5 days.

One final point – treat the answers that the software gives you like your change at the post office; check it carefully before leaving. The software is very accurate but very stupid. If you say that you have 12 painters and the software calculates that you need 12.3 painters for one day in June, it would decide that your figure is a failure and delay the project to eliminate the fault. Very few planners, if any at all, are so accurate that they need to worry about half a painter for a day in 6 months' time. We are trying to predict the future here; planning just is not that accurate.

Another common cause of error is to mix up your resource allocations. If you mean to allocate 40 man-days to a task as a work content and, in error, calculate 40 men per day, your answers are going to be miles off.

Many packages are unable to split a task into parts. Why in the world would you want to split a task? Here's why. Imagine you had a design job to do which was planned to take eight weeks and you also planned to attend a progress meeting after four weeks. You would create two tasks. One, called 'Design Task', would have a duration of eight weeks and the other, called 'Attend Progress Meeting', might have a duration of four hours. You would allocate yourself to both tasks and fix the meeting task in time with a constrained date (see Chapter 10) and expect things to work out well. No chance. Not if you try and level the plan.

Because the software cannot split the design task around the meeting it fixes the design task after the meeting. You are therefore planned to do nothing for four weeks, go to a progress meeting and then start the work.

When you are confident that your model makes sense, a neat trick is to move the plan's end date forward and backwards through time to see how this affects resource overdemands. Let's assume that you have imposed resource availability profiles for your major resources and imposed an end date on the plan. You might have 9 welders available to do work and want to finish the project by 20 October 2007.

You can play with the plan, adding some time here, taking away a resource there, moving the end date on a week. Each time you execute the resource levelling process. Repeat as required. This will give you an excellent idea of the balance between time restrictions and resource restrictions. You can even produce a graph of resource needs against time to impress your boss and also to help provide information on which a sensible decision can be made.

This graph shows, for a typical project, the relationship between overall duration and overall cost; that is, the total time it will take to execute the project and the total cost of doing it. By knowing the overall time scale and the demand for resources for each time scale

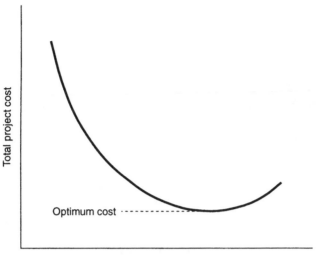

Total project duration

you can draw points on this graph. Fixed costs, such as project management overheads, are added in proportion to the duration. A long project means high fixed costs; a fast project means low fixed costs. The fast project may have higher fixed costs per day, but then there are fewer days.

The conclusion is that very fast projects cost a great deal, as they are very inefficient. Fighting a fire or a war is a very expensive, very fast project. You simply pour resources on to the project. They get used very inefficiently but the job gets done fast. As the planned overall project duration increases, the costs drop to an optimum level. After the optimum the contribution of the fixed costs causes the total project costs to rise once again. By examining such a graph, perhaps adding in the value of the completed project, you can make a very sensible decision on an optimal project duration.

Resource smoothing

You may also attempt to smooth the demand for each resource. Here you try to smooth out the jagged peaks and troughs of the histograms to improve resource utilization. You do this by adjusting the timing of activities within their float, by moving resources from activity to activity, and by many other techniques such as subcontracting and prefabrication.

It is worth noting that in engineering and construction, project managers are normally interested in keeping the demand for resources to a minimum. The resource pool is normally hired for the project. On shop floors and in information technology, electronics, and design offices, where the resource pool is stable and many projects go ahead at the same time, the driving force is to keep the existing resources fully utilized.

It is not wise to have 30 welders on your project in August and October and 3 in September. This may mean that 27 welders are sitting in the sun and being paid out of the project budget to get tanned. This sort of thing will make you very visible indeed. It would be much better to smooth out the demand for each resource in order that the demand is fairly constant throughout the project. This will probably be more efficient and much less expensive.

You can go too far with this smoothing. You can spend many happy hours fiddling away with your personal computer trying to find a neat, efficient way of achieving your project. I take a global view of resource smoothing – it is all very well getting the project into some form of shape, but there may well be a manager whose function is to decide who precisely does what job and when. If you spend too much time trying to do the manager's job, you will step firmly on the manager's toes.

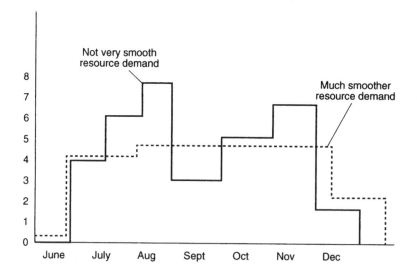

Also your resource smoothing may be thrown into confusion by a minor change to the plan caused by a very minor change in the weather.

Finally, while slagging off resource smoothing, the computer can only shift tasks about in time, trying to find better solutions. You, dear project manager, can do much better than the computer because you have a brain. You can do many things to improve the project's efficiency. Four suggestions follow.

Firstly, you can subcontract – this means using someone else's resources instead of your own. Subcontracting may seem more expensive and indeed may be more expensive, but it is a good way of overcoming resource shortfalls. Additionally, as the subcontractor will have some management input on the project, you effectively increase the size of the management team.

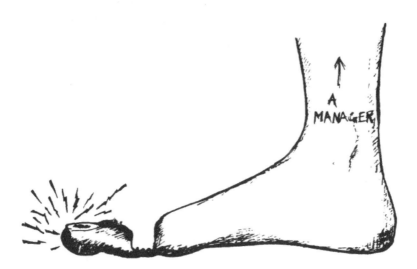

The manager's toes.

Secondly, take tasks out of the critical path and do them earlier. There are some items where work can go on earlier than normal, earlier than planned. By searching for tasks that could be taken out of their logical position in the network diagram you can often improve the resource-demand problems and find a faster way to run the project. The technical term for this is **prefabrication**. A little lateral thinking can save ages.

Thirdly, shift resources from task to task – now that you have the greater picture you can rethink your resource allocation, something which we said had to be done in isolation. There is an interesting argument that relates to this, and it goes like this.

The classic theory says that the critical path is the longest path through the network and is regarded by some as being inviolate – you leave the critical path alone and worry about the rest of the tasks. The counter argument says that the critical path is the longest path through the network because it joins up all the tasks which didn't get enough resources. Those tasks that didn't get enough resources tend to take longer and therefore tend to appear on the critical path. Therefore the first tasks to examine are those on the critical path.

By taking resources off tasks with float and moving these resources on to the tasks on the critical path you should find a faster way through the plan. This might be at the expense of increased risk as more tasks become critical or nearly critical, but the project's planned end date will be earlier.

Fourthly, check through your logic. Have you assumed that certain tasks must follow others when this really is not true? Are there better ways of executing the project? How about getting the team together and trying to find three ways of improving the project plan. You will get them thinking ahead about the project, and you will become visible as a problem avoider. Each proposal should be weighed in terms of benefits versus increased risk.

Resource hierarchies

In some project management systems, resources can imply the use of other resources. This is called a **resource hierarchy**. A typical hierarchy is as follows:

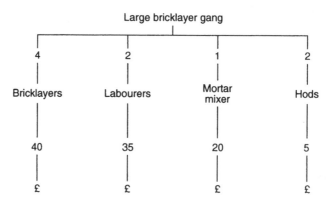

A resource hierarchy.

Notice how the resource *large bricklayer gang* assumes the use of 6 people and some equipment, and that the resource called *labourers* uses a resource called £ at a rate of £35 per day. If we use *large bricklayer gang* on a task, then the software will know that this implies the use of all those people and £260 per day. The lowest level in a resource hierarchy is often money.

You can use labourers on other tasks and in other hierarchies and, when you ask for a histogram of labourers, feel confident that they have all been included.

A hierarchy can be used to sum resources. Try this hierarchy on for size.

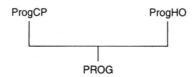

Here there are two types of computer programmer. One type is working in the company head office (ProgHO), and one is working on the customer's premises (ProgCP). The use of a ProgHO implies the use of a programmer (PROG), as does the use of a ProgCP.

The planner allocates quantities of the resource ProgHO to some tasks, and quantities of the ProgCP to others. The software adds these and calculates the total demand for programmers. Hence we can see the demand for three resources – ProgHO, ProgCP, and the total of these two – PROG.

A similar idea can be used to add up currencies, using the currency exchange rate in the hierarchy. To do this, you have a hierarchy which has the currencies you are using, each of which implies the use of a common currency. Expenditure in dollars might imply the use of expenditure in pounds times 0.75 where 0.75 is the currency exchange rate. As you allocate resources and costs to tasks in both currencies, some tasks using up dollars in the US and others using up pounds in the UK, the software totals both dollars and pounds. If the exchange rate changes, you have one number to alter – the number in the resource hierarchy.

Resource calendars

In many computer systems each resource can have a calendar associated with it. Such a calendar indicates on which days and hours the resource can work. No task can proceed on a day that is a holiday for any resource needed on that day. If the task uses resources such as analysts and programmers, then it can only proceed on days when analysts and programmers are both working.

In some software packages this is different from a resource availability profile. Sometimes these two concepts are connected. Tasks can usually be shown as proceeding on a working day even though there are insufficient resources available – the system tells you there is a problem.

Money management

Notice the absence of the standard term **cost control**. What we shall be talking about here is cost planning. Later we shall look at cost monitoring. Monitoring is to do with finding out what is happening. Control is doing something about it. Cost planning involves predicting the way money will be spent on the project in order that people can decide whether to proceed or not, and in order that they can arrange to have the cash ready at the right time.

There are some very handy tools to help you understand what is happening in the world of money but very little to help you take any sort of positive action. The tools put you in the picture; your brain decides what to do. So while people talk about **cost planning** or **cost control**, they mostly mean **cost monitoring**.

Money is a special resource. That folding, crinkly green stuff gets blown up out of all proportion. Money is rarely a problem; it's the absence of money that is a problem. Negative money is something to worry about.

Somehow in project management, banknotes metamorphose into small green bugs with long legs and strong desires to exercise those legs by disappearing over the next hill. Not only that, they are able to move in exactly the right numbers and at exactly the right speed. They manage to find exactly the speed which maximizes their population in migration but stays just below your level of perception. As many of the bugs that can will get away without your noticing until they are long gone.

Here are some tricks of the trade that will help keep the project's cash under control. You can start quite simply in the general area of cost control and get very sophisticated with money. Money is a resource, an important one but still a resource. The first simple level is to calculate how much each task needs of the money resource (dollars, pounds, euros?) and produce cash flow curves predicting the expenditure of money on the project.

In this simple case you say that a task will cost £1000 a day to keep going, or perhaps that a task will cost £12 000 in total. You get a planned running total of expenditure for all tasks. This simply means that for each day in the proposed project the amount of money planned to be spent on each task is added up. Cash flow curves look like those shown on p. 93.

Moving up a level in sophistication, you can use a project management system and tell it how much each resource costs. Then when you track your resources you will automatically be tracking money.

You can, of course, mix these two techniques. You could track important resources, each of which has a cash allocation, and allocate money directly for the other costs.

All this leads to one or more cash flow curves, most of which seem to look like a lazy S-shape. This will impress the socks off the project's founders because it tells them how much they will need to find and when. This is a real chance to become very visible in front of some very important people.

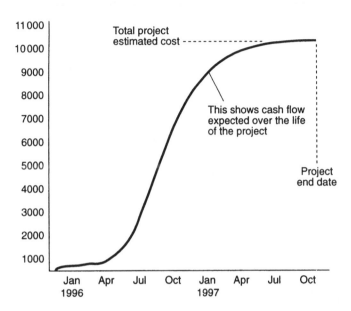

Of course the likelihood of the actual expenditure's bearing more than a passing resemblance to your cash flow curve is slightly lower than that of a three-legged cart-horse winning the Derby. However, by tracking actual expenditure we can provide some valuable services. A cash flow curve based on a plan for the project is called a **planned expenditure** and this can come in two float-dependent forms.

You will recall that many tasks have floats – the amount of time a task can be delayed without affecting the end of the project. If we assume that all tasks start as early as they can, we shall get one cash flow curve. If we then assume that all tasks start as late as they can, we shall get a different cash flow curve. The totals for both graphs will be the same. They will both start at zero on day one and meet at the planned total expenditure on the planned last date of the project. A graph showing both of these is called a **cash flow envelope**.

Financially speaking, we do not mind if the tasks happen early or late as long as they stay within their floats. Hence, we could draw a cash flow envelope, stick it up on the wall, and then, each month, plot the actual amount of money spent. If the line showing actual expenditure stays within the envelope, all appears well. If the actual line approaches or crosses the edge of the envelope, we have a problem.

We shall take a look at monitoring costs in the next chapter.

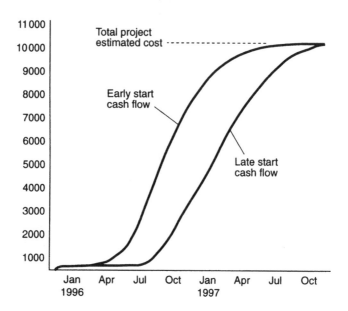

The process of incurring costs deserves a little thought. Do you mean paying the bills, placing the order, paying wages, or employing the staff? You see what I mean, I hope, that it is worthwhile thinking about what you actually mean by spending money.

Many people use 'cash commitment', an expression which ignores how long the accounts department takes to pay the bill but concentrates on what you have committed the company to pay. The amount of your company's commitment may be open to some doubt, as contractors often feel that you owe them more than you think. Arguments often occur about the amount of payment due for some work performed.

On the other hand you could deal with actual payments made – this avoids most problems caused by arguments, but costs take a great deal longer to establish. Some people combine both approaches, refining actual costs as more information comes to light. Your situation and the way your company works will probably decide this question, but do make sure that your planned and actual costs are calculated on the same basis. Compare, as my fruit-growing uncle used to say, apples with apples.

Progress monitoring and control

In this chapter the author discusses ways and means of measuring the actual progress of the project, and introduces concepts such as cost control and earned value analysis and finally imparts a few extremely neat and tidy ideas for the use of project managers.

Concepts

We have already stressed the need for a regular review and monitor process. Frankly, if you don't set up a regular progress and review session, you may as well not plan at all.

Project planning in the preproject phase, before work gets started, is a little like planting the seed, if you will forgive the rather biblical analogy. Planting the seed is quite hard work and in itself has some limited value, but not, as Paul Daniels would say, a lot. The harvest that the visible and successful project manager reaps is in the regular updating of the plan.

It doesn't take long. The hardest task is normally finding out what happened on the project during the last week. To begin the progress monitoring process, you will normally evaluate what work has been done on the project since the last monitor. This may involve a number of visits to the project and may mean discussions with contractors working both on and off the project.

There are two ways of approaching progress monitoring: DIY or OPE.* In the DIY approach, you do it yourself. You wander around the project measuring and counting to find out what work has gone on this week. You may find this very useful, as you are kept in touch with the various aspects of the job and the people working on it. While looking around the project you may notice little things going on, and, because of your relatively outside viewpoint, you may be able to come up with good ideas, which you can carefully suggest to the people responsible for doing the work.

> *These do not sound like recognized technical terms.

OPE (Other People's Effort) is the alternative. You get others to measure progress and report their findings back to you. There may be

some people whose job it is to measure progress for some other purpose – sometimes the work is measured so that people can get paid. Bonus systems and contractors' payment systems may provide useful input on progress.

The slight problem here is that other people measure the work using quantities that are not helpful to you. You need to know what has happened to each of the tasks in your plan. It is very interesting to know that the programmers have written 400 lines of code or that the bricklayers have laid 4000 bricks, but you must identify the walls that the bricks were laid in, or the program modules the lines of code were written for.

If no one is measuring what is going on, you could design some kind of a form and send it out regularly. This form lists the tasks that might be involved and leaves some spare space for comments. You ask everyone to fill up the form with their view of progress and send it back to you.

NORFOLK AND GOODE plc

PROGRESS AND MONITOR REPORT FORM

Project: New Advert   Date of Report: 10th June

Prepared by: Anne Uther

Task Number	Task Description	Planned Start I Finish		Actual Start I Finish		Expected Duration
12	design layout	12 may	12 jun			
15	prepare budget	5 jun	12 jun			
22	book space	12 may	10 jun			

It is normal to assess the actual progress achieved on a short list of tasks – those tasks which could be in progress. Progress can be measured as a remaining duration, an expected completion date, or a percentage completed. Estimating remaining duration – the amount of time it is thought that the activity will take to complete – is a much safer approach, as people tend to be more realistic, and no assumptions are made about the original estimates. If you ask someone how much of the work is done, they will reply with something like this: 'We must be 60% through that, Jim.' There are a number of problems with this statement. Firstly, your name may not be Jim, which would be very confusing.

Secondly, the temptation is to calculate that, as the task's original duration was 10 days and 6 days' work has been done, there must be 4 days left to go. This assumes that the 10 days you originally estimated was a reasonable estimate. Perhaps the task is going to take much longer, but was started earlier and is actually going to go on for at least two more weeks. Perhaps it was started only yesterday and will be finished tomorrow afternoon.

Thirdly, the human mind is hopelessly optimistic about measuring how far a task has progressed. Even more optimistic than it was in estimating the duration in the first place.

Over a period of time I have monitored the way people monitor progress. The estimated percentage completed starts off at zero. As the weeks roll by and work gets done, as a result of optimism this estimate increases more rapidly than the work itself. Soon the team realize that they have overestimated these percentage figures and then realize that they cannot go back. Reporting a figure less than last week's would show negative progress. Thus begins a lengthy period when the percentage completed figure is close to, and even approaches, but never actually reaches, 100%. This is referred to as persistent 99% complete syndrome and results in the saying that '99% of tasks in 99% of projects are 99% complete for 99% of the time'.

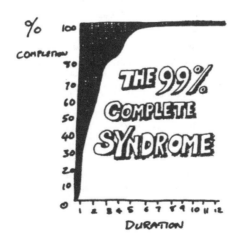

It is much better to ask when the manager thinks the task will be completed, or to ask how many days' work remain. The **remaining duration** is a good measure, as it does not assume a thing about the accuracy of the first estimate and nothing about when the task actually started. You essentially ask for the latest best estimate of how long the task will take.

You must have seen weeks where five days have passed, and, at the end of the week, the project has gone backwards, never mind forwards. Some helpful person has built a wall in the wrong place, and now, one week later, the new wall has to be taken down, while the right wall still has to be built. This sort of thing needs to be represented in the update.

This updating process can be carried out frequently throughout the life of the project. The frequency of the update cycle varies from project to project and during the life of each project. Fast-track projects at times of intense activity clearly require more frequent monitoring than do gentle projects in their very early stages.

You can reduce these problems of measuring progress by having tasks with original durations roughly the same as the update cycle. If a task has an estimated duration of 18 months, measuring progress once every month is going to be tricky. If it is broken down into four or five tasks of about 1 or 2 months each, progress reporting will be easier and more realistic.

One very important point. Many people make the assumption that because for 5 days people have been working on their tasks, 5 days' work must have been done. This is extremely dangerous. Did the people really do 5 days work, was it useful work, and was it spent on the right tasks? Just because eight designers have been beavering away on the design for the new boat does not mean that they have done much useful. They might have been painting the drawing office, or working on another boat, or they may have to throw away the work they have done and start again. Frequently time is spent overcoming obstacles or finishing other projects.

When the progress is established on a task-by-task basis, this information is entered into the project planning software, and the network is reanalysed, giving new timing for the remainder of the project, including a new estimate of project completion. This is not a huge job. On an average project there might be a few hundred tasks and around 20 or 30 active at any one update. Therefore you tell the software what day it is and update only those tasks on which some progress has been achieved.

The whole plan is reanalysed on the basis of the **update date** (normally the date of the update) and the amount of work that remains to be done. A new planned end date may emerge. New critical paths may emerge. You can then print new barcharts in time for the regular project review meeting.

A key item of data in this updating cycle is today's date. Anything that can go on can obviously start no earlier than today. It is really hard to start something yesterday. If, however, you do find a way, many project managers would like to hear about it.

Recognizing that the base date is not always today and recognizing that project management experts like complicated words where simple ones will do, we may mention a few terms used to describe this date. It is the date on which the software will start the remaining work. Terms such as **base date** and **time now date** are very common. The time now date will move slowly forward during the life of the project. This sketch shows the relationship between time now date, history, and the future.

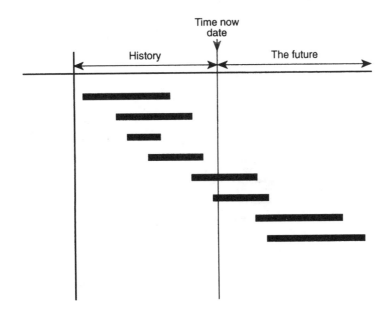

You may choose not to use today as the base date for the calculation of the plan. It may be more convenient to base the calculations on next Monday or the day after tomorrow. It would be unusual if this base date was more than a few days into the future, as too many things could change between now and then.

The people who sell computers talk about artificial intelligence and how clever the machine is, but half of these devices don't even know what day it is. You must choose a date of update and make it clear when getting update information that you want to know how long each task will take to complete as of that date. You ask questions such as 'On Monday, how long do you think will be left on the painting activity?'

This new analysis of the remaining work creates a new timing for the remaining tasks and therefore a new barchart. This new barchart can be compared with the original barchart on a task-by-task basis to see which types of work tend to be behind programme, and what the predicted effect will be of delays.

> What is the opposite of lateness?

The original barchart shows the proposed timing of the project as it was when you got the green light to go ahead with the job, at the conception stage of the project. Such an original plan is called the **baseline** or the **target plan**. It is your original time budget. A barchart comparing the current plan with the target plan is called a **slippage report**. Note the suggestion in the name that things are likely to be behind schedule.

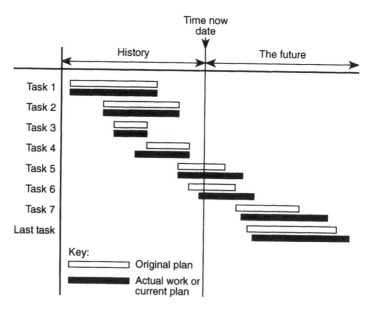

Slippage report.

Normally the project team would get a copy of the new barchart to enable them to predict when their involvement will be required. It is a fact of life that projects are delayed, and this in itself need not be due to bad management. It is, however, bad management not to know how much the project has been delayed and what the effects are. Can you see how people hiding delays out of fear or shame do great harm to projects? If they would admit to the problem, there would be time to find a solution. If they keep quiet and hope that their lateness will not matter or that the problem will go away, the time for problem solving disappears.

Of course, without a baseline or target plan, no monitoring is possible. Hence, 25% of the purpose of project planning is in having a yardstick with which to measure progress.

Bringing up problems early is an opportunity to become visible and solve problems, especially when the problems are not of your making.

An important part of progress monitoring is to understand what **issues** are currently making like difficult for the project team. If you understand what barriers are holding up the work, you can help overcome them. You might escalate these issues to the more senior managers. See Chapter 13 for more about risks and issues.

Cost monitoring

You will recall that in the last chapter we looked at drawing up cash flow envelopes in order to predict the way cash would be spent on the project. Let's assume for a rash moment that the financial wizards within your organization were asleep, or, even more unlikely, feeling happy, and they let the project go ahead on the basis of your plans and cash flow predictions.

To keep them both asleep and happy, you need to tell them how the actual costs are met regularly. You have a budget for each task, and, by adding up all the tasks, you have a cash flow envelope. You now need to track actual expenditure and compare the actual flow of money with the budget. This is monitoring – finding out what is happening.

Many software systems allow these sorts of data to be stored for each task. For each task you can have the planned costs (known as the **budget costs**) and the actual costs (known as the **actual costs**). The software compares these two and calculates the difference (known as **variance**). If a specific task is in progress – perhaps it is 33% complete – the software works out how much that task will cost to complete (known as the **cost to complete**).

All the system does is look at the way you said the task would eat up money and calculate how much the remaining part of the task should therefore need. If you estimated that a task would cost £12 000 and that task is 33% complete, the software would estimate £8000 left to spend. The software then prints out the three bits of cost information – **original budget, actual cost to date**, and **planned cost to complete**. It does this for each task and usually provides a sum total.

Before you can do any of this, you must, of course, be preparing regular progress reports – regularly updating the plan and informing the software what work has actually been done and how much was actually spent on achieving that work. Variance is a term used to indicate the difference between the original planned costs and the current planned costs of each task.

$$\text{variance} = \text{original budget} - (\text{actual cost} + \text{cost to complete})$$

This can be expressed as a percentage

$$\text{variance} = \frac{\text{original budget} - (\text{actual cost} + \text{cost to complete})}{\text{original budget}} \times 100$$

Some software packages allow you to control the sequence of printing tasks on reports. A clever trick is to print out the tasks so that those with the greatest variance come at the top of the report.

If you notice that certain groups of tasks are frequently overrunning the budget, you should examine these tasks and see what is going wrong. If certain types of tasks often come in well under the budget, you should also check into this to see if you can improve your estimating methods, or whether you should be promoting some very efficient manager.

Another part of the cash flow monitoring process is to plot the actual spend on the cash flow envelope. You find out what has actually been spent on the project at a specific date and plot that point on the graph. You could keep on doing this, joining up the dots as you go and watching the line of actual spend to see if it stays within the cash flow envelope. If the line showing actual spend stays within the envelope, things look OK. Going below the envelope shows underspend, and going over the envelope shows overspend.

This may all sound very good, but there are a number of significant problems that need to be addressed. The first is to do with interpretation. Imagine that you have been so good at running projects that you have become a project director. One of your team brings you a cost report for a project. (See figure on p. 103.)

Is this good news or bad news? Perhaps this project manager has found all sorts of cheap ways of doing the tasks so far and is expecting to arrive at the end of the project having made a substantial saving. Maybe the project is way behind schedule, not enough work has been done so far, and therefore you can expect to finish miles behind schedule. The only thing that the graph shows is that the project is not going to plan – it is either better or worse but we don't know which.

The graph provides no useful information. Actually a project could be behind schedule and spending too much to achieve what little had

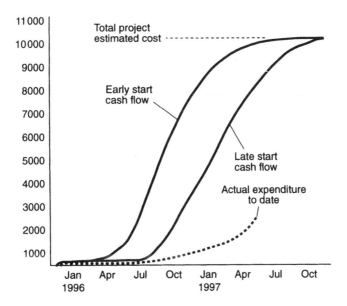

been achieved such that actual costs would appear to be bang on target. Taking a look at the project's barchart and comparing actual progress with planned progress might help to clear things up, but what we do not have here is a clever way of monitoring the project.

The second problem with this approach is the amount of time it takes to establish actual costs. If you are blessed with the normal accounts department, the chances are that it takes 6 to 8 weeks before it tells you how much you are actually spending. Information lags about 2 months behind actual expenditure. It can be worse than that. What happens is that over a long period of time, often much longer than the project's life, there is a growing certainty about the cost of each item. Let's take a single item which we plan to contract out.

In the beginning we have a rough estimate. This is later confirmed by an expert. Then we ask for quotes and the price becomes more firm. Then the work is done, and various arguments about technical specification, about who was providing support facilities, and so on, break out and are resolved. Eventually the contractor sends his invoice, which is too high and is argued about a bit. The invoice is paid but the guarantee period is just beginning. Finally, at the end of the guarantee period, the cost of the item is 100% firm. Perhaps that is the worst case that could be chosen for an example, but, as the fisherman said, you get my drift.

Often the monitoring function takes so long that control is impossible. By the time you find out what the problem is, it is too late to do anything about it. Some people approach this problem by dealing in cash commitment. This tries to show not what actually has been spent but how much has been committed – 'committed' means ordered, called off, or whatever. It is important to define the terms with which you are dealing.

What we really need is a system that somehow tells the true story in a simple and quick way. This is partly for our own information, but also so that we can report to the boss, the client, and the powers that be in general.

Enter **earned value analysis** (EVA) right on cue. EVA suggests that you compare the value of the physical work that you have done with the value of the work that you should have done. You measure the actual amount of work that has been done at a given date and multiply those amounts by the value – the cost rates used to produce the budget. You do not need to know how much has been spent, just the amount of work done and the value of that work. You can do this for each task and for total groups of tasks, and also total all the tasks for a project overview.

'Eva right on cue.'

This approach has many advantages. Firstly, you have to measure only the physical amount of work done – in many industries this is fairly easy. You are also comparing apples with apples – like terms are being measured. It is fast and a single piece of paper can show how the project is going. EVA even weights task in accordance with their cost. More expensive, more valuable tasks have a greater effect than cheaper ones.

You can add actual costs and do some useful, simple maths to arrive at various ratios each of which gives some idea of the state of the project. Keeping track of these figures gives you the ability to understand and report on the project's health and to watch for trends as they develop.

As a matter of interest, you need not think in terms of hard cash when doing EVA. As you are comparing like terms with like terms, you are free to use what terms you wish. There is no reason why you should not use worker-hours, cash commitment, numbers of bricks laid, or some other measure you find useful.

You would be wise to allow for indirect costs in an EVA. Indirect costs include project management, security, and other fixed costs. If the project costs provide for one project manager at £3000 per month and you are spending £4000 per month for a high-flyer, this should be represented in the EVA.

If a task is complete or not yet started, its value is fairly easy to compute. Tasks in progress are a little more contentious – how much have we earned by doing half of the pipe laying? This problem is magnified if there are many long tasks in the plan; therefore, EVA is made easier by having rather more short tasks in the plan.

Recognizing this approach, the US Department of Defense set up the Cost/Schedule Control System Criteria and uses this on most of its projects. The Department insists that contractors and subcontractors use certain approaches on defence projects, and this includes what we now know about EVA.

Here are simple explanations of the factors and ratios used in EVA. The meanings are explained in simple terms, as you will probably have a computer to work out the numbers for you.

These terms differ on each side of the Atlantic. In the UK we tend to use Earned Value Analysis (EVA) as opposed to the US Earned Value Management (EVM). Other key terms vary also, for each of the elements listed below, the UK term is followed by the US term:

BCWS – budgeted cost of work scheduled
PV – planned value

This is the value of the work you should have done at a given point in time. This takes the work planned to have been done and the budget for each task, telling you what portion of the budget you planned to have used.

BCWP – budgeted cost of work performed
EV – earned value

This is the value of the work you have done at a point in time. This takes the work that has been done and the budget for each task, telling you what portion of the budget you ought to have used to achieve that.

ACWP – actual cost of work performed
AC – actual cost

This is the actual cost of the work done.

SVAR – schedule variance
SV – schedule variance

This is the value of the work you have done minus the value of the work you should have done (BCWP-BCWS).

CVAR – cost variance
CV – cost variance

This is the budgeted cost of work done to date minus the actual cost of the work done to date (BCWP-ACWP). A negative CVAR shows the current budget overrun.

The following diagram shows a project which is currently behind schedule and overrunning costs.

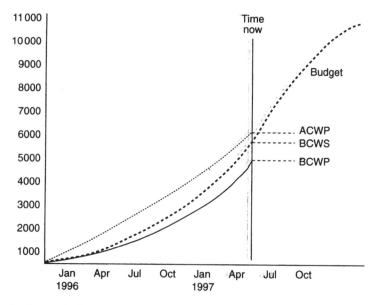

Earned value analysis.

EC (estimated cost at completion)

This is a revised prediction of how much the project will cost (taking into account what you have actually spent so far and your current estimate to complete the remaining portion of work).

BC (budget at completion)

This is the original budget for the whole project.

EVAC (estimated value at completion)

This is the difference between your original budget and your latest revised budget (BC-EC). A negative figure would indicate an anticipated cost overrun.

Finally you can calculate an index to provide a simple number to represent the state of your project

CPI – cost performance index = BCWP/ACWP
SPI – schedule performance index = BCWP/BCWS

In both cases an index value greater than one indicates a healthy project.

Reporting

Progress monitoring, cash flow monitoring, and earned value analysis are of very limited value unless you tell other people about them. This is one of those golden opportunities to make your presence felt, appear confident and in control, and generally give your favourite trumpet a quick toot.

On the other hand if the project is deep in the excrement, this might be a very good time to hide in the toilet.

Let's take a look at the major report formats that you are likely to use and explain their advantages and disadvantages.

Barcharts

These show very clearly the activities and their timing. (See figure on p. 77.) Barcharts can also show float for each task. Usually the critical activities are highlighted in some manner. With most software packages the displayed period of time and time scale (weekly/daily) may be controlled. Normally, this does not show logic or resource information. A barchart does give a strong clear overview of the project timing.

Barcharts may compare actual work done plus the current plan for work yet to be done with the original plan. Some people ensure that float of all kinds is hidden on their barcharts, as they take the view that if you tell anyone how much float he or she has, it will be used. In the extreme case, the planner disguises the critical path so the team cannot even tell what is critical and what is not.

Barcharts can have the logical links or dependencies shown on them. Such reports are called **logic linked barcharts** and these work very well on some projects. On other projects the sheer mass of logic links zipping about all over the barchart confuses the reader and masks the information you were trying to show in the first place. Some project planning systems work by displaying a barchart on the screen and allowing you to enter links between the bars. This again works very nicely, and such packages naturally lend themselves to producing printed, logic-linked barcharts.

Barcharts give a nice overview of the project or perhaps the next few months of it but fall short on accurate information. Barcharts do not normally show the latest start of tasks. The bar starts at an early starting date and reaches to an early end date with float shown, often with a different character, running out to a late end date.

Activity list

In text form this report lists the tasks, and prints for each associated dates (early start, late finish, etc.). While this report is more precise than a barchart, it does not give an overall impression of the project. When showing task durations and relationships, this report forms a useful way of checking data input.

Most software packages allow you to combine a barchart with an activity list on one report. The left-hand part is usually a table of task descriptions, start and end dates, resources and costs. The right-hand part is the barchart.

Task description	Early start	Late start	Early finish	Late finish
START	16/Jul/96	16/Jul/96	16/Jul/96	16/Jul/96
Site set up	16/Jul/96	16/Jul/96	20/Jul/96	20/Jul/96
Erect hoardings	23/Jul/96	23/Jul/96	25/Jul/96	25/Jul/96
Excavate foundation	23/Jul/96	2/Aug/96	26/Jul/96	7/Aug/96
Lay drains	26/Jul/96	26/Jul/96	3/Aug/96	3/Aug/96
Concrete foundation	8/Aug/96	8/Aug/96	17/Aug/96	17/Aug/96
Brickwork 1st floor	20/Aug/96	20/Aug/96	4/Sep/96	4/Sep/96
Windows 1st floor	28/Aug/96	7/Sep/96	12/Sep/96	26/Sep/96
Plaster 1st floor	31/Aug/96	12/Sep/96	25/Sep/96	4/Oct/96
Brickwork 2nd floor	5/Sep/96	5/Sep/96	20/Sep/96	20/Sep/96
Erect 2nd floor	21/Sep/96	21/Sep/96	1/Oct/96	1/Oct/96
Roof carpentry	21/Sep/96	10/Oct/96	8/Oct/96	25/Oct/96
Window 2nd floor	2/Oct/96	2/Oct/96	12/Oct/96	24/Oct/96
Plaster 2nd floor	5/Oct/96	5/Oct/96	30/Oct/96	30/Oct/96
Roof finishes	9/Oct/96	26/Oct/96	18/Oct/96	6/Nov/96
Watertight	19/Oct/96	7/Nov/96	19/Oct/96	7/Nov/96
Services 1st floor	7/Nov/96	7/Nov/96	22/Nov/96	22/Nov/96
Decorate 1st floor	19/Nov/96	27/Nov/96	30/Nov/96	10/Dec/96
Services 2nd floor	23/Nov/96	23/Nov/96	11/Dec/96	11/Dec/96
Decorate 2nd floor	4/Dec/96	11/Dec/96	24/Dec/96	24/Dec/96
Clean down	25/Dec/96	25/Dec/96	31/Dec/96	31/Dec/96
Handover	1/Jan/97	1/Jan/97	9/Jan/97	9/Jan/97
FINISH	10/Jan/97	10/Jan/97	10/Jan/97	10/Jan/97

You can design these combined reports choosing what to show in the table and what to show in the barchart to suit each team member and please almost everyone.

Network diagram

This shows the activities and their logical relationships. (See figure on p. 72.) The main use of a network diagram is to help the planner evaluate the plan. The question the planner asks is 'Does this software have a plan that looks anything like the plan I have in my head?' Typing errors, overlooked implications, and software bugs can all cause the plan in the computer to look very different from what you had in mind.

In terms of communication, the network analysis can be used to discuss fine points of logic with members of the project management team. Also the visible planner sticks his network plan upon the wall so that people can admire it and pretend that they understand what it is supposed to show.

What it really is supposed to show is that you are a superb asset to the company. Hence the network diagram can be neatly rolled into the shape of a trumpet and firmly blown. Some people achieve a similar effect by plotting out the network in colour and pinning it up on their office wall. Large plans may need pinning up in the conference room. Perhaps the corridor wall is the only one large enough for your super plan. It should not be displayed in a glass frame as it is very likely to change and need replacing from time to time.

Most network diagrams show no timing at all, but activity bars can be drawn to a time scale and have their logic links shown. This is called a **time scale network**.

Histogram

This shows the demand for a specific resource over time. Histograms may compare required resources with available resources and highlight expected problems that are due to underavailability.

Selective reports

Most activity-related reports – barcharts and activity lists, for example – can become long and unreadable to project managers and other members of the team. For this reason the reports are often shortened to show only a short period of time (for example, the next 3 months) or a

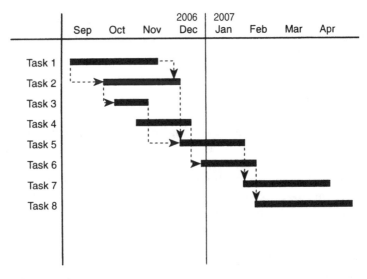

Time scale network.

selected group of activities. The **work breakdown structure** is especially useful in this respect as it allows the coding of the activities into groups. On this basis the operator can request activities to appear on the report that meet certain requirements. Hence barcharts showing only the design tasks, only the construction activities, or only phase 1 elements may be produced. Short barcharts covering the activities relevant to the reader maximize the communication aspect of the report.

To endear yourself to the project team members and do the project a favour at the same time, ask each member of the team what kind of report he or she would like. Then produce a report showing that information. Be flexible and keep checking that the report meets their needs. Tell them that it is important to you that they are kept informed about the project. It is true, after all.

Milestone reports

A special version of a selective report, these reports show only a management summary of the project by refering to the timing of a few selected key dates or milestones throughout the project.

	Week number												
	1	2	3	4	5	6	7	8	9	10	11	12	13
Start	O												
Approve project			O										
Acquire land				O									
Start to build						O							
Roof on										O			
Open for trade												O	

Milestone report.

Here are three simple little ideas that might help you keep people in touch with the project's progress.

Idea no. 1

This is Ron's report because to the best of my belief it was created by one who goes by that name. It applies where there is a number of tasks within a section of the work, and you would like a graph showing how those tasks are going. It is particularly nice if the tasks run through the same basic process. Let's take, for example, an office-carpeting project. In each room we have to clean the floor, purchase the carpet, and lay it.

The vertical scale could be square metres or money, so that the height of each bar relates to the others in a consistent way. The number of tasks has to be small to make this work (2 or 3), and there has to be some degree of repetition.

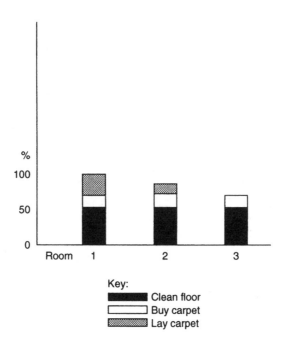

Idea no. 2

The Z-form

This applies to a fairly long, repetitive task. Cleaning or replacing 2000 windows would do. The rate of progress is what we want to know about and to track. The whole thing might be represented by the graph below. The line marked *plan* shows the rate at which we originally planned to do this task. The line marked *actual* represents what has happened so far. Pretty bad. The line marked *recovery* shows the rate of production we would need to achieve to get back on schedule and finish the task on the originally planned completion date. The line marked *go on* shows what will happen if we continue at the current rate and therefore how late the task will finish.

This idea is limited to tasks where the rate of production is fairly constant, but it does smack people right between the eyes and drive the message home in a succinct way.

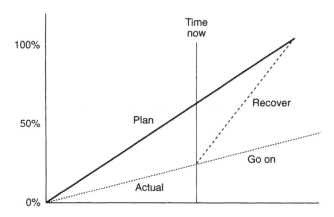

Idea no. 3

One day the project will be finished. Everyone will be delighted it is over, either because it has been a huge success or perhaps because it has been a miserable failure. Don't rush off and forget all about it. Do a post-project report. This is particularly useful if there is a likelihood of similar projects being undertaken in the future.

In organizations where many projects are being carried out, the project team analyse their historical projects to arrive at a better basis for estimating future projects. How much did the project actually cost? How long did the tasks actually take? Real experts build up valuable databases of output levels for craftspeople, machines, and other resources which will enable future project managers to estimate more accurately the next projects.

Prince2, governance and

methodologies

This chapter introduces the concept of governance through a formal project management methodology and discusses the utilization of the more popular tools and especially Prince2. Examples taken from common methodologies add colour and will benefit the reader. Some warnings about the misuse of these tools are imparted.

The term 'methodology' is popular but inaccurate. The 'ology' suffix should refer to the study of a topic; hence zoology, psychology and pharmacology.

So consider this question: what is the difference between a project management method and a project management methodology? The answer, dear reader, is simply the price. The theory says that because the name is long and sounds complicated, more organizations will be happy to pay larger sums for a purpose-built, complex sounding methodology than they would for a plain old method.

This book will firmly stick to the word 'method' as a small voice of protest.

A method is a set of rules or guidelines

To explain a method let's venture out onto a football pitch.

Imagine a green pitch, immaculately trimmed by caring grounds-men, where 22 people will spend the next 90 minutes playing a beautiful game under the watchful eye of a referee, two assistant referees and thousands of gentle, placid, adoring fans.

There are rules and guidelines by which this game is to be played. One rule states that only the two goalkeepers and referees are allowed to handle the ball. Another rule defines the duration of the game, how it will begin and what will happen if there is a draw after the allotted time. Guidelines explain how one player may tackle another and what will happen when unfair tackles are made.

Now David Beckham, Wayne Rooney, Thierry Henry and I all understand these rules. If understanding the rules was enough, we would all be great football players and I would have a Ferrari. It is a fact that you could be an expert in the rules of football and have never kicked a ball. This describes many fans.

Clearly understanding the rules is not enough: Messrs Beckham, Rooney and Henry are able to play the game magnificently, using their skills to make decisive moves and passes, spin the ball in the air and dodge round other players.

So it is with a method. A method gives you the rules and guidelines by which projects should be run. It defines how they should start, what happens at key stages, what responsibilities each person has and how it all should end. Understanding these rules does not make you a great project manager. You could be an expert on a method and have never taken part in a single project.

Unfortunately many organizations expect their managers to go off on a Prince2 training workshop and come back transformed into expert project managers. This is not going to happen. Prince2 trainees will understand the rules but their ability to run a project will be unchanged by the Prince2 workshop.

However, if your organization has a number of project managers running a number of projects, there will be value in having them all follow some consistent approaches. A common method will give the senior management confidence that the project workload is being consistently and professionally managed. The adoption of a method can be expensive but it does allow the organization to demonstrate a high level of governance.

So your team needs to understand your method and learn how to manage projects.

Prince2 is by far the most popular project management method in use within the UK. It was written by the UK government through the Office of Government Commerce. You can attend training workshops offered by a range of accredited training organizations, sit an examination and become Prince2 accredited at one of three levels. Its slightly odd name is an acronym abbreviating **PR**ojects **IN** a Controlled Environment. It is now in its second form.

An alternative to Prince2 is published by the Project Management Institute from their headquarters in Philadelphia. Their method for managing projects is referred to as the PMBoK – the **P**roject **M**anagement **B**ody **of K**nowledge. PMBoK is more a set of guidelines rather than rules, but it is popular throughout the world and has been interpreted into complete methods many times. Again, accredited training providers will help you to qualify through examinations in PMBoK.

Another popular method is called Pride. This stands for **PR**ojects **In D**eep **E**xcrement but it is not to be relied on.

There are many other methods. Some are based around PMBoK or Prince2. Some are home-brewed within a single company or organization. The major consultancies have their own methods which they will be happy to amend to suit your organization before rolling it out to your team and mailing you significant invoices.

To help you understand the content of a formal method, we will now pull back the curtains and outline the concepts of Prince2. This is a very, very, very brief summary of Prince2 and it will only give you an idea of what it contains.

Stages

Prince2 suggests that you break down each project into **stages**. Each stage begins with a specific set of deliverables, products, documents and approvals and ends with the deliverables, products and documents ready for approval that the stage was designed to deliver. The function of the stage is to progress from the products at the start of the set to those at the end. Each stage ends with check point at which the project is reviewed before approval is given to proceed to the next stage.

Sometimes the handover from one stage to the next involves a new stage manager and you would be wise to accept a project in these conditions only when you are satisfied that everything is in shape and hunky-dory. Taking over a stage in a method-driven company is much like buying a second-hand car – you should look before you leap.

Prince2 does not say what the stages should be as this will depend on the organization and the project, only that there should be stages.

Organization

The section on organization within the Prince2 method explains how a project organization should be structured and who is responsible for various aspects of the project.

Prince2 discusses a range of roles that should be adopted. Often these roles are part-time and often one person may take multiple roles on a single small project. In a large organization running a major project, these roles can be carried out by a whole department.

For example, Prince2 recommends the establishment of a project board. The board manages the project from a fairly high level. Three roles exist on the project board apart from the obvious project manager:

- Executive   corporate guidance and assessment
- Senior user   representing the users of the project's deliverables

- Senior technical user   represents those responsible for the
technical aspects of the project.

Prince2 suggests there may be two project management roles. The project manager plans the project, reports on progress to the board and initiates activities. A stage manager may be appointed for each stage of the project (design stage, installation stage, etc.) and is responsible for delivering the products of that stage and meeting quality, time and cost targets.

There should be, in the world according to Prince2, a product assurance team. Made up of a business assurance co-ordinator, a technical assurance co-ordinator and the user assurance co-ordinator, this group has two responsibilities. One is to ensure that the project board can believe what the project manager says about the project and the other is to offer help and the benefits of their experience to the project manager.

Prince2 also recommends the formation of a project support office where the experts in programme management, project management and project planning can sit and share horror stories about the project workload. The business assurance co-ordinator and technical assurance co-ordinator roles can become specialist roles performed on all projects in this centralized project support office. Representing users is likely to be more project-dependant.

Resource conflicts

Where a number of projects are in hand within an organization, there is nearly always conflict amongst the project teams for precious resources. Teams fight for the resources they need to complete their project, but do not have the benefit of the broader view of the work. The broader view permits understanding of priorities between the various projects.

To help in this respect, some companies establish project boards whose role is to act as a referee between the various project teams. Project boards expect a standardized form of reporting from all projects managers. Every project manager reports to the project board. When questions of priority arise, the project board is aware of all the workloads and conflicts and is able to make decisions about resource allocation.

Project boards may also set and maintain standards with respect to project management methods, bringing a degree of standardization to the project culture.

Plans

This section talks about the planning documents that Prince2 users prepare or expect to be prepared. Prince2 talks about plans of various kinds and you may experience the difficulty I found when they use the word 'plan'. When I think of a plan, I get a mental picture of a barchart or Gantt chart or perhaps a PERTchart. Some of the Prince2 plans are not likely to look like a barchart at all.

Plans start with the technical plans. These are traditional time-scale plans showing activities that will go towards achieving the project goals.

Resource plans are used to identify the number of resources and the type of resources required to achieve the project. The project resource plan is a top level plan for the whole project and the stage resource plan is a more detailed look at the resource requirements for a stage. These will look much like histograms in graphic or tabular form. Resources in Prince2 can be human, equipment and other resources required or cost.

The Prince2 guides talk about appropriate levels of planning. There can be, for example, within the overall project technical plan, the following plans:

- a stage technical plan for each stage in the development process
- a detailed technical plan examining in greater detail some complex or major activity
- individual work plans detailing specific people's activities within a stage
- an exception plan is produced when things have gone wrong, showing what action is to be taken to deal with the deviation.

Other plans can be broken down in a similar way.

Prince2 is keen on product-orientated planning. A key feature of the Prince2 philosophy is to focus on products rather than tasks. The idea is that you concentrate on the deliverables of the project and the deliverables of each stage within the project. This focuses people's thinking towards the products of the project and the parts of those products that need to be achieved by each phase or stage.

The danger is that once tasks and small jobs are identified, it is easy to concentrate on achieving those tasks and losing sight of the larger picture in terms of what you are actually trying to achieve.

To develop this concept further, Prince2 talks about the Product Breakdown Structure (PBS) which is much like a Work Breakdown Structure (WBS) but concentrates heavily on what is to be produced.

Controls

There are two forms of control within Prince2.

One deals with the documents that the project creates. These documents need to be controlled so that the right people get the right version of the right document at the right time. The other side to controls within Prince2 is about the methods that the organization uses to keep a beady eye on the projects and to propose control actions.

Controls in Prince2-speak are either a meeting or an assessment and we can look at three types of meetings and two assessments.

The project initiation meeting is designed to be the point of final approval to go ahead with the project, the moment of commitment. Of course the meeting could not sensibly assemble all the data in one session so someone has the job of assembling a Project Initiation Document which contains loads of information about the objectives, timescales, budgets and aims of the project.

I am very keen on this stage as so many projects get going with only the vaguest idea of what they are supposed to be aiming to achieve. More importantly each person will have a different idea and it is very likely that your boss's idea of what you are supposed to be doing is likely to be different from your own. Therefore no matter how good a job you do, your boss will feel a little miffed. It is unfair but true. A solid definition of what the project is about and a formal signing-off will reduce the chances of this happening. It will not reduce them to zero.

As the project gets going it passes into its first phase which might be a design phase or a detailed requirements definition stage. You might consider a mid-stage assessment to provide a control point for the project management and project board. At such an assessment the team compares progress and development with the original aspirations and makes decisions. If things are going very badly the project might be shelved or scrapped. If things are going well the team might give the go-ahead for the next stage even though the current stage has not ended.

More often the project is ambling along reasonably normally and the team talk about the unplanned situations that have arisen and make decisions about dealing with them.

During each stage the team are likely to meet regularly at a checkpoint meeting. These are regular opportunities to review who has been doing what on the project in the last few weeks. At each checkpoint meeting the stage manager, helped by the business assurance co-ordinator and technical assurance co-ordinator, produces a checkpoint report for the project manager.

At the end of each stage there is likely to be an end stage

assessment. Here the project managers present to the project board a report defining the current status of the project. They seek the board's approval to proceed to the next stage. Now get that idea stuck in your head. At the end of each phase or stage of the project, the project manager has to prove the case to proceed with the project. It is a great way to keep those (us) project management people on their (our) toes.

Finally there is a project closure meeting which closes the project and hands the deliverables over to the users. The meeting is supposed to devote time to examining the history of the project and learning lessons from it that can be applied to later projects. The team are generally far too busy sinking pints of real ale or sipping gin and tonics to celebrate the end of another project to worry about looking back.

That is Prince2 in a nutshell. The actual set of documents is voluminous as it goes into detail about the agendas for meetings, descriptions of the various roles and the contents of each report. You can take Prince2 out of the box or adapt it to meet your own needs. Adapting it can mean simply changing the names and titles used to suit those within your own organization. It can mean a thorough re-write. The definition of the stages is very likely to need tuning to suit your lifestyle.

If your organization does not have a method, perhaps you might suggest that you investigate the subject more fully and then run a project designing and installing a method appropriate to the company. If this unlocks doors with government, you might enhance your personal reputation a touch.

It is worth thinking about who is best suited to creating a method for you.

You have a number of choices here. You can recruit full-time or part-time or consultancy people to come in, learn about your company and write a method just for you. Your own personal 'ology'. Of course you will have to pay them for their time while they learn about your organization and the way it works. You'll also have to pay your own staff for the time they spend talking to the consultants explaining how the company works.

Another approach is to take some people within your own organization who are familiar with its systems and methods and get them to learn about method theory and create one for you. Rather than take a methods expert and teach them about your company, you take a company expert and teach them about methods.

Perhaps the ideal is a mixed team of both. One of the more risky approaches is to select a fine, mature employee whose current prime objective is to survive quietly until the retirement party. As this person ran some really great projects in their hey-day and since they know the company inside out, they should be ideal for the job. Their motivation is low and they may find it difficult to understand what this method is actually for.

Some organizations take a bright young thing who has recently joined the organization. Said bright young thing devotes some time to researching with a fresh eye how the organization works and makes notes in a procedures manual or a method. There is a spin-off benefit here apart from the method itself and that is you get a bright young newcomer who is familiar with all aspects of the company's machinations.

Methods tend to get stuck in a groove. As time changes, as organizational structures change and as the company metamorphoses to react to those changes, the methods have to be kept up to date. There is a danger that the method becomes the holy grail, the organizational equivalent of the law. People often blindly obey the structures of the method long after it has become irrelevant to the company's current way of working. It pays to set up a group now and then to update the method in light of changing circumstances. I have never seen a company with a suggestions box for improving the method but some system for keeping an eye on the method reduces the danger of stagnation.

Some final words

Prince2 is a toolbox and, like all good toolboxes, it contains a set of tools, each of which suits a different need. In some organizations and especially within the public sector it is often assumed that all tools must be used on all projects. It is not unknown for the cost of applying Prince2 controls to more than double the cost of the project and this is clearly ludicrous. This is a mistake and not what Prince2 is designed to achieve. Prince2 should be applied sensitively and sensibly.

I very much support the idea of having layered versions of your method. In one organization I helped establish three forms of one method, graded for simple projects, standard projects and, in its most demanding form, for complex projects. We even established a process for selecting how a method should be selected based on the size, complexity, sensitivity and risk of the forthcoming project.

Here is a simple PERTchart showing a simple process flow for a simple project.

You can see how the Project Initiation Document (PID) is prepared and submitted to the project board. Once approval is given the project has only two stages. The first stage might cover the design activities and the tendering process. This first stage ends when tenders from suppliers have been received. At this point an end-stage assessment takes place. After approval the second and last stage proceeds where the orders are placed and the contractors get on with the work.

Finally the project ends with project closure.

This does not mention any of the activities required to actually manage the project, only those required to control it. Remember that methods are about the rules of the game, not the expertise required to actually play.

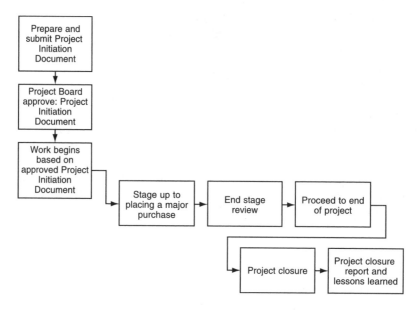

As much of Prince2 relates to documentation, it makes sense to link the required project control document to the project control tasks in your plan. Here is a simple project plan template listing the major process steps and referring to the appropriate documentation for each element.

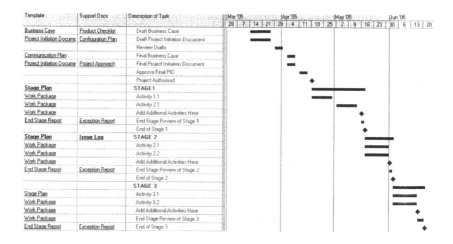

Advanced critical-path topics

The assiduous reader has, by now, absorbed the basics of critical path analysis, but there remains much that can be explored to advantage. In this chapter the author outlines certain more advanced topics that extend the usefulness of the critical path technique.

In our tour through the project management toolbox there are a few rather specialized tools tucked away in the smaller drawers. We shall look at some of the topical and popular techniques that spring from an organized approach to project planning. Generally, these topics assume that there is a network model somewhere for the project, but that could be just a list of tasks. To take advantage of these techniques, you will, almost exclusively, have to load a model of the project on to a small computer.

Work breakdown structure and organizational breakdown structure

One meaning of a work breakdown structure

Work breakdown structure (WBS) and **organizational breakdown structure** (OBS) are methods of grouping activities by a coding system. They are very fancy names for a simple idea. The problem that these methods try to solve is volume. No, not the knob on the TV – but the number of tasks in the project. It is not unusual to have 300 or 400 tasks in a project, and 1000 are not uncommon. There is the odd project with 10000 tasks, but these are mercifully few.

Now, if you are going to produce reports for a few different people every week, and these reports are each going to have 400 tasks in them, you are going to use paper very fast. Moreover, the recipients of your reports are going to have to wade through all this paper trying to find what concerns them. They won't bother. Too much paper destroys communication almost as fast as too little.

Wouldn't it be nice if you could get the computer to extract certain groups of tasks in order that recipients get a short report telling them about the tasks they are interested in. Computers are supposed to be great at that kind of thing. These managers would have to study only the bits that concern them, but you can remind them that you are looking after everybody's tasks and the complex interrelationships among them. This is another opportunity for you to blow that old trumpet of yours in a subtle way.

WBS and OBS give you exactly such a coding system (you're not surprised?). What is special about WBS and OBS is that the coding system is structured. Usually you have some space attached to each task which you can use to create a coding system. You normally allocate specific spaces or character positions to do certain jobs, creating a multidimensional coding system.

Here is an example. This is a new factory we have to design and build in which we are going to print books about project management.

Type of task	Work breakdown structure
Architectural tasks	AR----------
structural engineering tasks	SE----------
environmental tasks	EN----------
design tasks	---D--------
construction tasks	---C--------
main building tasks	-----MB-----
car-park tasks	-----CP-----
structure tasks	-------ST---
fit-out tasks	-------FO---

Thus, if you want a list of tasks for the main-building structure, you would ask for those tasks with a work breakdown structure such as this: -----MBST---. If you wanted architectural tasks involved with the design of the car-park fit out, then you want a WBS such as this: AR-D-CPFO---. Finally, if you want all the design work in the fitting out, you want tasks with a WBS such as this: ---D---FO---.

In an OBS, the concept is very much the same except that the categories relate to departments in the organization rather than subdivisions within the project. Departments are divisions such as the design office, sales, marketing, import, export, personnel, and so on.

Another meaning of a work breakdown structure

In some software packages and indeed in some business environments, WBS and OBS take on an extra depth. The story goes something like this. It is very important in project management that everyone knows who is doing what to whom and with what. The division of responsibility is the key. If you do not have a clear idea of responsibility, if you do not assign the tasks effectively, people will be confused. Some jobs are done twice, both times with people who think that they should be in charge of this particular task.

Even worse, some jobs are not done at all. When the next project meeting is held, someone will ask if the quotes are in for the printing, and everyone will look at everyone else. A short, acrimonious argument will follow during which people will deny that they had the responsibility for doing the work, people will carefully reread the minutes of the last meeting, and accusations of incompetence will be made freely. None of this gets the project moved forward, but it does emphasize the need to have a clear allocation of duties.

A WBS can help here as it is possible to build a hierarchy or outline structure showing how the tasks fall into groups, how the groups fall into groups, and who is responsible for each group. In projects where time is not a key factor, the project management team may never get past this stage. All they need is a clear allocation of work.

The allocation of tasks into groups, the creation of a WBS, need take no notice whatsoever of time. The WBS is like a different dimension cutting at right angles across the time dimension. Tasks can appear within a WBS and a PERTchart, all within the same computer model. Some systems draw out the WBS for you.

Here is a WBS showing how a group divided the workload in arranging an office move. One way of achieving this family tree of job allocations is by means of an outliner, and indeed some packages do offer this facility. The outliner originates in word processing when you start with a document which is to be structured into sections. First headings might be chapter headings, and under each of these come section headings. Sections may be broken down into subsections and so on. This results in paragraphs with numbers such as 1.2.7, which is the seventh subsection in section 2 of Chapter 1.

Using an outliner, you can collapse sections within their headings to see an overview as well as inspect documents word by word. This can be done section by section, allowing a detailed look at one section, with a summary of the others.

ID	Task name	November 23	30	6	13	20	27	December 4	11	18	25	January 1	8	15	22	February 29	5	12	19
1	The Project								Summarizes whole project										
2	Sub-heading A								Summarizes phase A										
3	Task 1																		
4	Task 2																		
5	Task 3																		
6	Sub-heading B								Summarizes phase B										
7	Task 1																		
8	Task 2																		
9	Task 3																		
10																			

Now, we can apply these concepts to project management. We can create a plan, starting with main sections of the work. Then we can break down each section into subsections, break these subsections down into sub-subsections, and so on. If we want, we can see the whole plan in all its glorious detail, or we can collapse some sections to show only a heading level but leave the detail of one section on show. The way in which tasks are grouped into headings and subsections should have no effect on the critical path or the timing; it is a matter of convenience and the ability to concentrate a report on a certain aspect of the project.

Summarizing the whole project by collapsing all tasks into their headings often produces the report that ideally suits the managing director. The range of other ways depends on the software package you use, but here is a case study, taken from *Project Manager Today* magazine in which a structured plan was the order of the day.

Work breakdown structure.

Case Study: Universiade XVI – the 16th World Student Games

'The biggest sporting event ever to have taken place in the UK' – that is the 16th World Student Games. All of the competitors are supposed to be students of one form or other, and many of the athletes will go on to compete in the only bigger sporting event – the Olympics.

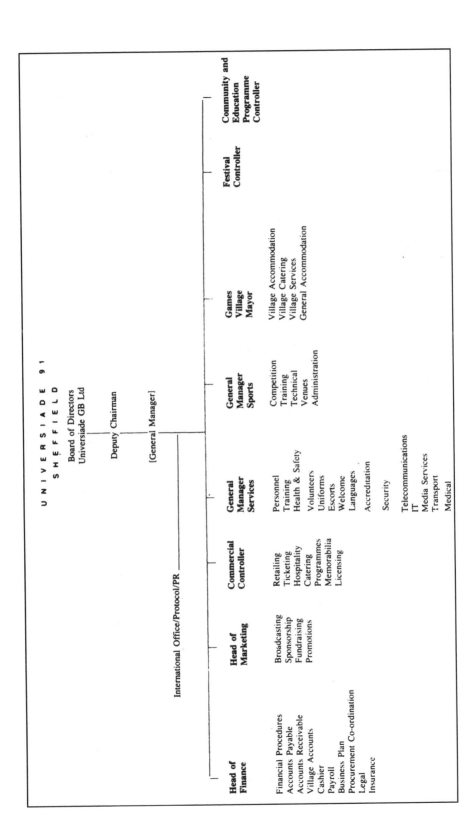

Now, when you think of a major sporting occasion, your first thoughts probably go one of two ways. If you are a builder, you think of those stadia (the correct plural according to my dictionary) and swimming pools. If you are an IT professional, you think of scoring and timing systems. The games involve many buildings and lots of computers, and that is just the beginning.

Take a look at the management chart of Universiade Great Britain Limited.

UGBL is the company which was formed to organize and run the games. You will see the major departments into which the organization is split as well as a short list of the subprojects within the major project. From this you may get some idea of the project's scope.

Let's take catering as an example. There were 6000 competitors and officials from 130 countries living in Sheffield for the two weeks of the games. They all had to be fed in accordance with their national and religious needs as well as their special needs as athletes. They had three nutritious, muscle-building meals a day. Not so with the VIPs attending the sporting conference, the judges and umpires, and the huge number of team-support members – they tucked into more up-market fare. There was guzzling going on at the student village, most venues, the conference, and the VIP facilities. The total grub requirement was 40 tonnes per day.

The caterers didn't even start with a kitchen. In the build-up period they had to organize the kitchens and cafeteria, bars, and buffets. It is not surprising that UGBL had a full-time catering expert seconded from Gardner Merchant – the kitchen equipment company.

There were five purpose-built venues – this building work was organized by Sheffield City Council – and in these venues the hopeful athletes grunted and groaned their way towards gold medals. The village to house all of these visitors was converted from the semi-derelict Hyde Park Flats in Sheffield, most of which will be put back into the local housing stock after the games. There was one of those spectacular opening ceremonies. There were 998 competitions – including 192 athletic events, 216 basketball and 88 hockey matches, and 180 swimming races. As this is a friendly competition, the losers do not simply drop out but take part in runner-up contests, so that every visitor got to play, run, jump, dive, or whatever quite a few times.

The venues overspilt from Sheffield into South Yorkshire, causing a real transport problem. There had to be a fleet of vehicles to take the competitors from the village to the right location, as well as to move the VIPs and judges, not to mention the transport of the huge number of visitors who came to watch their hero, heroine, niece, or boyfriend compete.

The venues could not be completed, commissioned, and immediately pressed into use for the games. They needed realistic testing as venues.

Hence, most of the new venues were made available to local clubs to use in the period between building completion and the opening ceremony of the games.

Everyone attending the games had to be vetted to ensure security. Every name was entered into a purpose-built software system, so that when participants arrived they could be issued with suitable badges to allow them into the relevant areas. This is called accreditation, which was only a small part of the security subproject.

Add medical, drug, and sex testing, media facilities, and a daily news-paper, and you get some feel for the size of the project. As if that were not enough, Sheffield threw in a cultural festival to compete with the Edinburgh Festival, and every group in the town from the cricket club to the tiddly-wink society did something to coincide with the main games.

Planning

While the games were planned at a broad-brush level by network plan-ning, and barcharts existed for most of the elements in the project, this work was put on hold following some management changes. One of the problems with the planning even at a summary level was to create a work breakdown structure (WBS) that would suit the current requirements and yet leave room for the unknown future needs of the project.

A WBS is a structured system for coding tasks by relating them to various groups, phases, management responsibilities, or other groupings.

UNIVERSIADE (GB) LIMITED

PROPOSED WORK BREAKDOWN STRUCTURE   JULY 1989

FUNCTIONAL DEPARTMENT	DIVISION	CODING SYSTEM
OPERATIONS	ACCOMMODATION	A *
	OPERATIONS SUPPORT	O *
	TECHNICAL SERVICES	T *
	SPORTS	S *
	PROTOCOL	P *
MARKETING	SPONSORSHIP	S *
	PRESS OFFICE	P *
	MERCHANDISING	M *
	CEREMONIES	C *
	EDITOR	E *
	INFORMATION SERVICES	I *
FINANCE	INFORMATION TECHNOLOGY	I *
	PURCHASING	P *
	ACCOUNTING	A *
	FINANCIAL CONTROL	F *
	OFFICE MANAGEMENT	O *
HUMAN RESOURCES	VOLUNTEERS	V *

Commercial software packages handle this in different ways – some are very structured and some have a very free format. The various functions within the organization suited the management needs but bore little relationship to the natural subprojects within the project itself. Hence, the system shown on p. 129 was developed.

The software used permits a 12-character work breakdown code. The first character was devoted to the operations function, with *A* = accommodation, *S* = sports, and so on. The second character was left free for later subdivision within these categories. The third character indicated a marketing involvement with a *C* = ceremonies and an *I* = information services. A *V* in the seventh position indicated a need for volunteers.

In addition to this, the tasks were numbered in groups: 12100 to 12199 set aside for accommodation at the student village, 12200 to 12299 for accommodation elsewhere, and 14100 to 14199 for shuttle transport. The task numbering relates to the subprojects, whereas the work breakdown relates to management functions. The tasks can have only one number but could have many characters in the WBS. The table

UNIVERSIADE (GB) LIMITED

PROPOSED TASK NUMBERING SYSTEM   JULY 1989

SUB-PROJECT STRUCTURE

1ST LEVEL	2ND LEVEL	3RD LEVEL	TASK NUMBER RANGE

1ST LEVEL	2ND LEVEL	3RD LEVEL	TASK NUMBER RANGE
SPORTS	EVENTS	SCHEDULES	11100 TO 11199
		VENUES	11200 TO 11299
		EQUIPMENT	11300 TO 11399
		TICKETING	11400 TO 11499
	ACCOMODATION	VILLAGE	12100 TO 12199
		SUPPLEMENTARY	12200 TO 12299
	ACCREDITATION		13100 TO 13199
	TRANSPORT	SHUTTLE	14100 TO 14199
		PUBLIC TRANSPORT	14200 TO 14299
	TELECOMMUNICATIONS		15100 TO 15199
	INFORMATION TECHNOLOGY		16000 TO 16500
	VOLUNTEERS		17000 TO 17999
FESTIVAL	COMMUNITY FESTIVAL		21100 TO 21199
	COMMERCIAL FESTIVAL		22100 TO 22199
CEREMONIES	OPENING		31100 TO 31199
	CLOSING		32100 TO 32199
	ASSEMBLY		33100 TO 33199
	CONFERENCE		34100 TO 34199
PRODUCTS	COMMERCIAL		41100 TO 41199
	EDUCATIONAL		42100 TO 42199

shows the task numbering system. As the plan was built, the planners allocated each task within its task number range and entered any characters in the WBS that were relevant.

If you want a barchart for the sports department, you need to check for an *S* in the first column; a *C* in the third position indicates a task connected with the ceremonies. Any character at all in the fifth column indicates a finance connection. It is easy, therefore, to get a barchart for a part of the plan. Telling the software to look for an *S* in position 1 and a *V* in position 7 gives a barchart relevant to volunteers working on the sports subproject. A requirement for task numbers between 12000 and 12199 gives details of the student village, and tasks between 12000 and 12299 give all the accommodation activities. Asking for tasks numbered between 12000 and 12199, plus a *V* in the seventh position of the WBS, gives a barchart of those tasks in the student village that require volunteers.

By this means, a barchart can be produced for most needs, there is considerable scope for expansion of the plan within the structure, and most people can get a single A4 barchart for their area of interest. The accreditation barchart (overleaf) shows the timing of the subproject which is concerned with setting up a database of visitors in order that they can be checked on arrival and issued with badges giving them access to the appropriate areas.

Merging plans, boiler plating, or task processing

In most network analysis systems you can merge plans together to create a new plan. The command depends on the software system you are using, but there is usually some system for building one plan from components. Word processors do exactly this. You create and store documents on your disks and add those documents into other documents.

For example, you might have a number of files containing the addresses of all your major customers. You might also have a file containing the following words:

If you do not pay our outstanding invoice number _____, dated _____, we will send _____ large men round to rearrange your kneecaps.

Thus, when you want politely to remind clients that it is time they paid your last bill, you create a document, pull in their name and address, pull in the standard threat, and fill in the gaps. You do very much the same thing with little or large bits of network plans.

Merging has a number of uses. Let's suppose you are a senior planner in a company and have four planners running various projects around the country. Each project is planned separately, and each

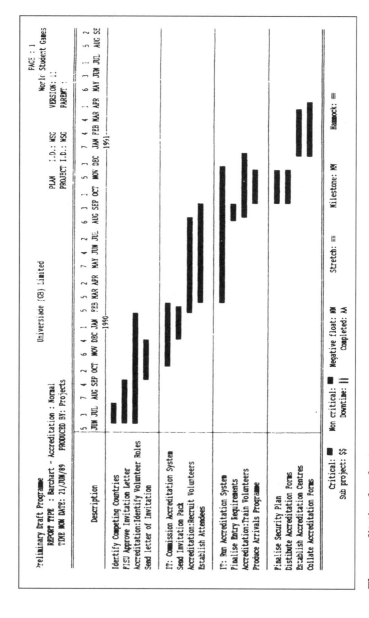

The accreditation barchart

planner tracks his or her own resources. From time to time, it might be useful to collect all these four plans into one large plan to get a look at the company's cash flow on all projects. The ability to merge would be ideal for this, as you could get the planners to send in their latest plans, and you could merge them together. This is sometimes known as multiproject consolidation.

If you have to merge plans together regularly, you could store a special plan designed to link the various plans together. Perhaps your plans share resources which get passed from one project to another. Create a small file of tasks that in itself makes no sense but holds the logical connections that you need to link the plans to be merged. Let's call this plan **interface**. When you merge the working plans together and throw in the interface plan, hey presto, you have a fully linked, multiproject plan.

Merge is also useful to create plans that are repetitive. If you need a plan to build 20 similar marine gearboxes, you could create a typical gearbox plan and merge it into a global plan 20 times, making minor adjustments as necessary. Here is a little plan that is the sort of thing that you might keep on your disk so that you can pull it into your plans from time to time. It is called Acquire.

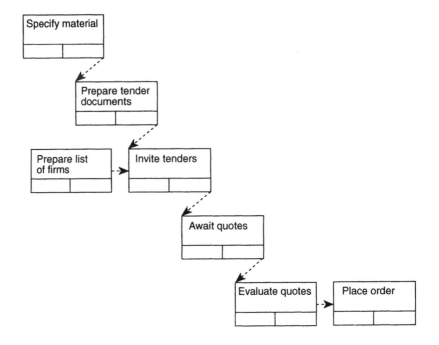

Linked projects

Some software packages allow the operator to 'link' projects together or to 'include' many projects into one. The purpose of linking projects together is to get a view of the overall resource requirements of the group of projects as a whole. As mentioned before, assuming that all projects are being planned in a similar way, it would be possible to evaluate the predicted demand for each resource on all projects and identify conflicts. This is very similar to merging plans together but rather more temporary.

Linking and merging are often done by companies who have many simultaneous plans leading to a programme management solution. Programme management refers to those companies in which many projects are going on at any one time and all of which call on the same precious resources. *Programme Management Demystified* (Reiss, 1996) deals with such topics more fully.

Hammocks

Hammocks are tasks for which you don't work out the duration. You link the hammock task into the plan at relevant points, but you do not specify a duration. The program calculates the timing of all non-hammock tasks, and then deduces the duration of the hammocks themselves. A hammock fills up all the time available between its starts and finishes.

Imagine two trees with a hammock hung between them. It's on a desert island, the sun is shining, there is a cool drink in your hand, and the surf is gently lapping at the golden beach. The hammock hangs between the two trees, exactly spanning the gap. OK, now wake up from your reverie, and lets get back to planning. That's why hammock tasks are so called.

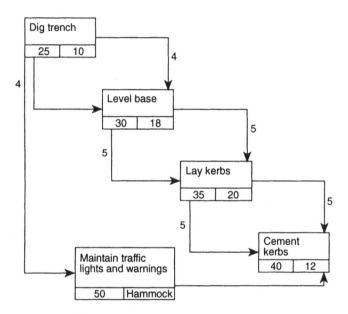

Hammocks are especially useful to hang resources on to. Your project probably has fixed overheads or running costs for the management, security, office space, and whatever. These are a fixed cost per week for the duration of the project, however long that is. Easy – tack the cost of these overheads as a resource to a hammock linked to the start node and end task of the plan. The project management software works out how long the project is and therefore how long the hammock is.

The cost of the overheads is multiplied by the newly calculated duration. If you find a quicker way to do the project, the hammock becomes shorter and the cost reduces. If the project lengthens, then the hammock gets longer and the fixed cost goes up. Hammocks can be used anywhere in any plan; therefore, if overheads go up for a part of the project, the use of more than one hammock will handle these steps.

There is another handy trick that you can do with a hammock. You can hang hammocks around parts of the project attaching the hammock to the start of the first few tasks and the end of the last few tasks in a group. What happens is that the hammock expands to fill the time available, exactly covering the time taken by the task group. Hence, it provides a neat summary of a group of more detailed tasks.

By clever use of a number of hammocks and by using WBS codes to select certain hammocks to be printed on a barchart, you can get a summary of the project in a few well-chosen lines. You can emulate a proper WBS system by the clever use of hammocks. If you make changes to the plan, you can feel confident that the hammocks are stretching and shrinking to fit the right amount of time.

Hammocks can have more than one start link and more than one end link. They should start at the same time as the first task and end at the same time as the last one. This is useful if you are not quite sure which is going to be the first or last task in a group. Some systems use the term 'span task' for hammocks.

Calendars

A calendar is a basic tool of the network planning engineer. Network analysis is a mathematical process – it might be calculated that a certain task is going to start 234 days into the project. This is not the sort of information that tells you a great deal. The 234th day needs to be translated into a date. To do this, the system must know which days you are working and which days are holidays. Given that, it can count 234 working days into the project and calculate a date such as 25 January 2007.

A calendar dictates which days you work and which days you rest. It also indicates which part of the day you plan to work. Calendars are separated into two components:

Work week: this defines the typical working week.
Non-work calendar: this defines special non-working days,
e.g. Easter, May Day

There may be any number of calendars, each of which will have two parts as above.

Activities and resources can be associated with a different calendar. You can simply work with one calendar for your plans and avoid this complexity. If, however, you have most tasks and resources working 5 normal days a week, but some special tasks can only proceed at the weekend, such as road surfacing, then these calendar functions may be useful.

Sometimes you may have many calendars associated with one plan. One plan might relate to work going on in the UK with Saturday,

Sunday, and Christmas Day as downtime. Another might relate to work on the project site in Saudi Arabia, where the weekend is on Thursday and Friday, and Eid is a holiday.

Network Rail has design tasks that can proceed only during the normal working week and track work that can happen only on Sunday mornings. With some systems you can have special calendars that relate to specific resources. Perhaps you can use the crane only between Tuesday and Friday each week.

The software package takes care of all these things following a set of rules that should be set out in the operating manual. You can take advantage of these advanced calendar capabilities in a number of ways. Firstly, every activity must be associated with a calendar. When you create an activity, you will be asked for its description, duration, and other factors including the activity calendar. Secondly, with most systems, you may link each resource to a calendar. When you define a resource, one of the questions will require you to link the resource to a calendar. If you do not specify a calendar, the project calendar will be used.

This does not mean that you must create a calendar for each activity and each resource. You might have only one calendar that applies to all activities and resources. You might have two calendars, of which one might apply to all activities in the docks and the other to all activities at sea. For any task to proceed on a given day, it must be a working day as dictated by the task calendar, and any assigned resources must be working as defined by the resources calendars.

Constraint dates

Normally, in project management, you tell the computer program about the activities and how they relate to each other, and it tells you when these activities should occur. There are times when you want to ensure that certain activities happen at a certain time. There are times when factors outside your control have an effect on your plans. These are also called **external restraints, must dates, not before dates** or **scheduled dates**. Examples of such external restraints to a builder or engineering contractor include the delivery of information from the design team.

Constrained starts and finishes override the normal critical path calculations on the plan. A constrained start means that an activity cannot start before a given date. A constrained finish means that a particular activity must be completed by a specific date. These constrained dates are considered to be very important, and the software will respect them wherever possible. You can build in a constrained start date which controls the part of the plan in question and call the constrained start 'receive design information'. This will indicate to the design team how vital to the project their work is.

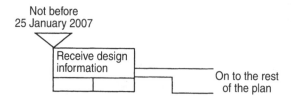

Another example of an external restraint is the delivery of an important item of equipment to the project. If you are responsible for the delivery of the equipment, then you should probably have a small subproject that concerns itself with the selection, inspection, and delivery of the piece of plant. If someone else is delivering it to your project and you have no control over it, you can represent its delivery by using a constrained start date at the part of the plan that is concerned with installing the piece of equipment. Constrained finishes represent occasions where a part of the project has to be completed before a given date, perhaps when you hand over a part of the project to someone else.

You may sprinkle a large number of constrained start and finish dates throughout the plan and control it through these means. Almost any activity may be constrained by tacking on to it a constrained start or finish. An official opening ceremony might be fixed in time by using both a constrained start and a finish for the relevant activity. By saying it cannot start before a date and that it must finish by a date, you are fixing it absolutely in time.

Negative floats

The use of constrained starts and finishes can give rise to curious phenomena called negative floats. Negative floats are bad news for the project. They draw your attention to planning problems.

Suppose that you have a part of a plan restrained by a constrained start and a constrained finish, and that there are 100 working days between these two dates. Also suppose that the activities in this part of the plan have a critical path that is 120 days long. This means that you are trying to do 120 days' worth of work in 100 days. Your software will report a negative float of 20 days on the critical path. There may be other activities with smaller negative floats, and there may be activities with positive floats.

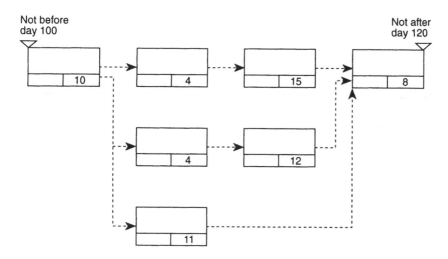

In the sketch above, in which only activity durations are shown, the main activity will have a negative float of 17. The activities total to 37 days, and yet only 20 days separates the constrained start and finish. These negative floats are problems; they should be removed by resequencing the work, reducing durations, or changing the constrained start or finish. Continuing to run a project with negative floats is like driving with a flat tyre – things need mending before they get worse.

There may be more than one negative-float path. You may have activities with −5, −3, and +5 days float. Removing 5 days in total from any of the tasks on the −5 path should resolve that problem but may leave the −3 problem. Rerun the network and evaluate. If the problem still remains, take at least 3 days out of any of the tasks on the −3 path by increasing resources, increasing overlap, and doing more activities in parallel, and then analyse the network once again. Eventually you will arrive at a zero float path showing you the critical path to an achievable project.

Some project planning software systems cannot deal with negative floats. During the analysis phase such systems hiccup and report an error before abandoning the calculation. At least this forces you to do something about it.

Subprojects

You can have many levels of subprojects in most network analysis systems. The idea of a subproject is to group tasks together into neat bundles. You may then treat these tasks as a group and represent them by a single task. This allows you to deal with your project at various levels – in great detail if you wish; in global terms at other times.

Let's imagine you are building a four-engined aeroplane. First you might create a subproject for an engine. This would include tasks involved with the engine components and assembly. You would use this subproject twice in the plan detailing the wing. This would show the spars and struts in the wing and show two tasks, each of which represents the engine assembly and fitting. Two such wing plans would be used in the main plan along with a fuselage plan, a tail plan, and a fit-out plan.

Thus, Subprojects allow you a convenient way to build up a complex plan from parts. The duration of a task that represents a subproject is calculated by analysing the PERTchart of tasks within that subproject.

One common use of subprojects is in a repetitive type of project. Imagine building 50 houses of two types. You would create and refine a plan for each house type. You would refine each house plan to optimize resource usage. You would then create an overall plan using 50 of these subprojects and optimize the total resource demands by sliding the representative task for each subproject around to find a neat, economical way to phase the houses in relationship to each other. This multi-tiered way of working with plans is called a **hierarchical** approach. In the hierarchy, plans get more detailed as you go down levels. Plans at one level include representations of plans at lower levels.

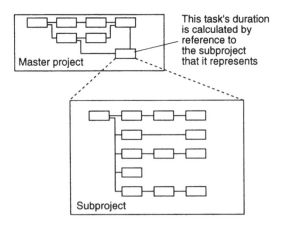

Beware – some systems do not permit the creation of links between differing subprojects. The subproject is represented at the next level up by a task, and that task can have links coming into it and going out of it – the representative task can have preceders and succeeders. You may not be allowed to create a link from a task within a subproject to another task within another subproject. Such links are called **off-project links**. If you need off-project links, take care how you buy your software package and how you create your subprojects.

Claims

Many project managers are involved in contractual situations that stem from competitive bids. In such cases, one company promises to perform a project for another company which, in turn, promises to pay for it. The company doing the project is called the contractor, and the company paying the bill is the client. This relationship exists in a huge range of industries from building to software, and from manufacturing machinery to printing. There is usually a third party that dictates the exact nature of the project – the professional team. This team might be architectural, engineering, or system consultants.

The very nature of the competitive-bid contract is likely to lead to arguments. These arguments usually start during a project and go on long after it is finished. The dispute is about who is going to pay whom and how much. Network planning is a useful tool on most projects because it helps predict problems and therefore helps overcome them. These programs lay foundations for a case should disputes begin. The planner can evaluate the true reasons for a delayed project. We shall discuss some techniques for using critical path analysis to substantiate dispute arguments.

The **as sold** *network*

It is a normal demand of most contracts that the contractor produce a programme showing how the work will be done. Increasingly, professional teams are demanding network plans from their contractors. The US Department of Defense has laid down very strict demands for its contractors, and this trend is spreading to many other industries and countries. We shall call these plans **as sold** plans. We shall assume that they have been submitted to the professional team and that no adverse comment has been made.

What if there is no network?
I recognize that in many cases in which disputes arise a network does not exist. The main contractor usually produces a fairly small barchart

showing rough timings for the major steps of the project but showing no logic whatsoever. I think that in such cases the main contractor is missing an opportunity to

1. help achieve the project more efficiently;
2. lay foundations for a procedure to resolve disputes should they arise.

I also think that it is valid to synthesize a network retrospectively. If a project has got into trouble and disputes have arisen, it is possible to examine the contractor's as sold barchart and information-required schedules, and to deduce a network from that information. Clearly, the case to be argued will be weakened, while the question of the validity of a synthesized network is a legal one which I cannot comment on as I know of no test case.

I shall assume that a network diagram exists. Either it has been prepared for the project, submitted to the architect, and not disapproved, or it is later synthesized. I also assume that this is done on a computer.

Laying the foundation of the arguments

There are four main topics in which disputes can be helped by the use of critical path analysis.

Information required

A very popular area – the professional team must produce information to complete the design process and in order that the contractor can get on with the work. Here a powerful tool is to build **acquisition subprojects** (ASPs) into your plan. Each acquisition subproject shows the flow of some information from the professional team to the contractor.

Tasks such as *receive pipe schedules* are the sort of thing I have in mind. This precedes a task that represents the work the contractor has to do to get ready to order the materials or negotiate with the subcontractor, i.e. tasks such as *take off materials*. Then comes a task which shows the actual ordering and negotiation of the materials – *negotiate steel pipes* would do nicely. Next in this little ASP comes the delivery time for the materials. Finally comes a task which shows the delivery of the goods, and this ties in to the main construction network at a suitable place – into the task that uses the material or subcontractor.

Such a network would look like this:

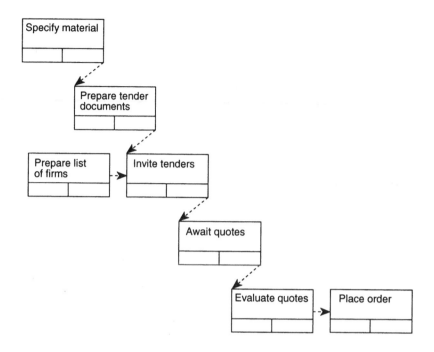

You could have many such ASPs, each referring to some element of design that entails information flow from the professional team.

To help the project along in practical terms, you will be able to take a listing or barchart showing the information-required dates and issue this to the designer or client. You will also be able to list the negotiate tasks for your buyer and the delivery dates for the project manager. If information is late, you will have a very strong case to show how the agreed plan indicates the effects of the delay. We shall soon discuss the **as built** network idea for the method of capitalizing on this.

Client control dates

Client control dates refer to any date imposed upon the project by the client or professional team. For example, certain areas of a building behind handed over to the contractor, access made available, equipment ordered by the client and due on a certain date and decisions that the client must make are all client control dates.

It makes sense to show these on the network diagram and to use the as built network technique to emphasize the effects of delays on these dates. It makes sense to arrange client control dates on or near the critical path. Remember that the critical path is not a magic thing impartially calculated by the software.

The critical path is the path where you did not put your resources.

The tasks which use few resources and therefore happen slowly are likely to be near the critical path. In a practical, project-planning role, and when a plan is analysed, you can start to take resources off tasks with float, slow those tasks down, and move the spare resources on to the critical tasks to speed them up. For the purpose of this discussion, then, we can assume that a competent planner can steer the critical path to some degree. Hence, put your client control dates on or near the critical path.

Nominated subcontractors

Where you do not have responsibility for the performance of a nominated subcontractor or any other organization doing work on the project at the same time as your own company, you should ask the designer for the programme of work agreed with the nominated subcontractor. You can build this into your plan. Once again it is useful to have any nominated subcontractors on or near the critical path.

Weak areas

Sometimes you feel that there are areas in the project where insufficient design work has been done, or for some other reason the nature of the work is uncertain. In construction, provisional sums and mechanical and electrical work are like this. You should ensure that this work is clearly identified with the network plan. Then, if the work does actually lead to variations, the effects will be easier to prove.

Some other points here – if you are very aggressive about disputes, it may be that you prepare one network for the submission stage and another for the actual work. If you are synthesizing a network retrospectively, you would very likely ensure that the points I have just made were handled.

Approval

Approval of the plan in these terms takes two forms. Generally, the design team will receive your plan, stick it in the bottom drawer, and promptly forget about it. This is fine. The basis of any future disputes is laid.

If you retrospectively create the plan, the approval stage is more complex. There could be legal argument about the validity of the exercise. More probably, there could be argument about the specific logic you had used. It could be argued that the contractor would not have used a particular link like that; he would have linked this task to that task. Perhaps the durations of the links would be disputed.

It is also very sensible to have one task showing how you accept the client control date, the nominated contractor's work, or the provisional sum work. There should be one filler task between the task with external constraints and those showing pure execution of the work.

As built networks

As built networks (ABN) are a method of using a plan created in network format to show how the actual delays affected the work. I must emphasize that this technique is not used to build a case to show a specific point – it is used to discover the truth. If you can discover the truth, I suggest that you will be in a better position to argue the case. You will know where your case is weak and where it is strong. Also these techniques are primarily aimed at evaluating delays, not consequential loss and expense.

The basis for an ABN should be an agreed network diagram. Ideally, this would be a network diagram which had used the tricks mentioned above and which had been submitted to the designer and approved as the contract programme. Hopefully, while the project manager was executing the project, good records of information flow were kept, including daily project records and records of delays which were due to all factors.

Now for the claims evaluation.

Complete hold-ups

Something that has completely stopped the work on the project must be dealt with first. If heavy snow stopped the project, or a shortage of cash has been encountered, you should change the calendar on which the network plan was processed to show the days you were unable to work as non-working days or holidays, as in the parlance of most project planning systems. Then the network diagram is reanalysed and printouts run off.

All days off which were due to factors that affected the whole project are allowed for in this way. You were unable to work on certain days, and, as the plan shows, these major delays are allowed for. Hence, you can see what would have happened if there had been just the major hold-ups and nothing else changed. Running the network through in this form would report that. This will give you a good idea of how important the total stoppages were on this project.

Specific delays

Specific delays are problems that delayed some part of the work. Delays that prevented some work from proceeding can be inserted by means of a constrained date. Problems that slowed work down, causing tasks to be lengthened, can be shown by increased durations, but a convenient way to do this is to add a task alongside the task in question and link it with a start-to-start link from the start and a finish-to-finish to the end of the affected task. The extra task could be called something like 'Delay due to change order', and the duration would be the actual amount of time it took to execute that task.

The process of analysing the effect of these specific delays goes like this. You take one of the early delays – let's say the late delivery of some piece of information. This is added into the as built network as a constrained date on the relevant task, showing that the task could not have started before a specific date. Rerun the plan and get a new end date showing when the project would have finished if the delayed information was the only change – all other tasks went as planned.

Then repeat the exercise with the next delay. Perhaps, at about this stage of the project, you brought in some changes to compensate for the delays caused by the design team. Add these changes – perhaps a reduced duration or an increased overlap. Reanalyse the plan and again get a new end date. You can repeat this analysis, each time reanalysing the plan and getting a new planned end date for the project.

This gives you a table which lists various problems, various delays and various attempts to regain time, each of which results in a new end date. It is likely that the last item on the table should give an end date very close to the actual end date of the project. The end dates can be plotted on a graph, and this reinforces statements such as the following:

> The original target end date was 10 March. The design information for the foundations came late. If no other changes had occurred, the project would have ended on 1 April. We tried to regain some lost time by prefabricating foundation shutters, and this brought the end date back to 14 March. But then the first-floor column layout came late, and this would have pushed the end date to 15 April. The delay in adjusting the gutter design did not change the end date, as the project was already delayed so much that the delayed information actually came in time.

What tends to happen is that the list concentrates the mind on the delays that really did delay or advance the project, and hence the wheat is separated from the chaff. If you want to present a very strong case, you should be very careful about the sequence of problems that you apply to the plan. Taking the problems in one sequence can cause very different answers to another sequence.

You can prepare two sets of runs on this basis – one with the original calendar and one with the bad weather recognized. This means that you can show how, assuming good weather, the delays would have caused one state of affairs, and how delays combined with the actual weather to cause disruption. Now you have a plan that examines the major stoppages, delays caused by external restraints on task start dates, and delays caused by change orders that affected task durations.

You can see that critical path analysis can be a powerful tool in evaluating delays and supporting a dispute argument. I believe that it is likely that you will find that only a small number of delays actually affected the work, and this may change the way in which your dispute cases are presented. The growing frequency of network plans demanded in contract documents will lead to more frequent use of network diagrams, which might lead to a rational approach to disputes.

The people issues

The simple fact that people run projects is herein acknowledged along with the recognition that a short chapter could not attempt to analyse fully the wide-ranging issues relating to teams of people. However, the people topics – motivation, team building, and delegation – are here introduced to the reader in order that those wishing to discover more can at least be guided in their search.

As Huckleberry Finn probably never said, 'I ain't seen no horses running a project.' Project management is about people.

There are two schools of thought about people in project management. One group believe that science and technology can aid the project manager in motivating and controlling a team. The other attitude is that dealing with people is a natural thing and that most people will motivate themselves if given half a chance. I am firmly in camp two. Most people, given half a chance, enjoy their work and get on with whatever they should be getting on with. All that is needed is for the manager to give them the opportunity to do exactly that.

Regrettably, many managers do not. They unwittingly tie the hands of their team behind their backs, they stick them in rotten situations where their talents are not appropriate, and then they do not take the time to listen to them and therefore do not realize that this is the case. They do not praise their staff, they give little thought to the career progression of their staff, and they do not reprimand their staff.

A reprimand can be a very positive thing.

Some excellent, thought-provoking work has been done in the field of management that will help even a half-witted project manager to recognize what is going on. We are going to talk about teams, people, motivation, without going completely over the top, getting all scientific and trying to get a Ph.D. out of the thesis.

I believe that in management, answers are easy; it is the questions that are hard to find. Most questions have a series of possible answers, and selecting the best answer is relatively easy. The hard bit is finding questions that need to be asked. So what I hope to do in this chapter is raise questions in your mind.

Management is nearly always a transaction. Most people both manage and are managed. Most people have people that they are responsible for and people that they are responsible to. Most people feel that their bosses are unreasonable and unfair, and should be put out to pasture as soon as possible. Many managers feel that their staff are lazy and good for nothing, and that they themselves could do a great job if only they had a better team. Good managers treat others as they themselves would like to be treated. But most managers treat others as badly as they themselves are treated.

Something we can do is to try to classify people into types in order to build a rounded team with people in suitable roles. We shall look at some methods of people classification. Thus, I hope you will find this chapter thought-provoking, and I hope that you gain an insight into managing people. We shall not attempt, in this short chapter, to cover this enormous topic thoroughly – we can only open up the can of beans and take a peek inside. I hope that you will be a good manager, make people happy, and achieve your projects. I hope your boss does the same.

The role of the project manager

Delegation, authority and responsibility

Delegation and authority are exchanged by managers and the managed. The manager assigns a task to a member of the team and delegates the authority to do whatever is necessary to achieve that task. The team member accepts that authority and takes responsibility for carrying out the work.

The manager still has responsibility for carrying out the work, as the manager chose the team member and issued the instructions. As a manager, you should always support your team. Just because someone reprimands you for doing something wrong, don't say that you gave the job to Henry and he bungled it. It is unfair to Henry, and you denigrate your ability to select the right people for the job. You should either accept the criticism or argue the case as if you did the job yourself, and then go and talk to Henry afterwards. You will command respect among your staff and your peers.

When people say that a job has been done badly, they can mean one of two things. One is that the job was done badly, and there are not many ways out of that. But very often they mean that the job was done not in the way they would have done it, but differently. There is a huge chasm between doing a job badly and doing it differently. A good manager accepts the way the job was done, takes responsibility for it even though he has delegated it, and perhaps points out some alternative strategies later on.

Leadership

Leadership becomes a problem in project management because projects nearly always involve a team of people who, frequently, have been assembled for a specific project. Leadership can be defined by the need for a leader to:

- define and achieve tasks – task needs;
- build up and co-ordinate a team – team needs;
- develop and satisfy the individual members – individual needs.

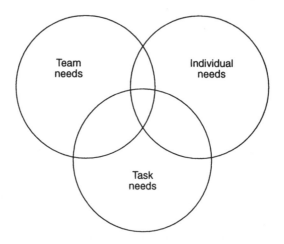

Let's summarize the main points. These three needs can be expressed as follows:

Task needs
This refers to a group's desire to achieve their objectives. Clearly, a project team have a need to achieve the projects that they set out to do, and leadership must aim towards these goals.

Team needs
This refers to that element of leadership that attempts to hold the group together. If the team members are working together in a co-ordinated fashion, the results of their work will be greater than the sum of the parts. Conflict and misunderstandings which will inevitably arise must be used effectively.

Individual needs

This refers to the leader's role in catering for the individual need of each team member to contribute to the project, to understand how well he or she is performing, to show his or her potential, and to be recognized for good work.

The leader must attempt to satisfy these three interrelated needs. The leader who concentrates only on the task – 'going for the project' – and overlooks the encouragement and motivation of the team may do well in the short term. Eventually, the team will give that leader less than their potential. A leader who concentrates only on creating team spirit may tend to neglect the project and will eventually gain a reputation for late projects. The previous diagram shows how every leader must consider these three overlapping needs. Go and make a cup of tea and consider the leader's role in two common situations – the captain of a cricket team and the conductor of an orchestra – in these terms.

In project management, the problem of leadership is a little more complex than in ordinary management. To begin with, very often the project team is thrown together at short notice. Because of the nature of project management, people often find themselves working on a project with a crew of strangers. There is little time to adjust to those co-workers; there is a job to get on with. In ordinary management environments, there is a turnover of staff, but the team tends to exist for a reasonably long period of time. Rough edges rub together and smooth each other off. The team works well together. From time to time, a new member arrives and causes some new friction, but eventually the team resolves its differences. Project management teams have no time for this.

There is another problem heightened in the project management arena. The project management team has a desire to achieve a specific objective. How do you motivate people when the end of the project is in sight? The nearer the project is to completion, the nearer the team members are to looking for a new job. It is relatively easy in the multi-project environment, in a company where there are 102 simultaneous projects going on, and where the completion of one project just means the start of another. But in big projects, projects where a team is striving to create some large new thing, motivation gets harder and harder. In particular, the individual needs are hard to handle in the big project arena. You want to give people a route to grow, a career progression that will challenge them and not overstretch them. But major changes, like the end of a project, make it very hard to deal with such things.

It is in the nature of project managers to create things. You find project managers building model cars or railways in their spare time. But the act of creation, be it a model railway, a bridge, a tunnel, or a new computer, is an anticlimactic thing. Project managers tend to lose interest when the project is nearly over.

In many organizations, people who are reasonably happy in their jobs are plucked out and made into project managers. Typically, someone who has caught the managing director's eye is selected to manage the launch of a new product or the move to new premises; a one-off project within an organization not used to projects. Is this the big break that is going to elevate you to new heights within the company, or is this going to be the biggest mistake of your life? Are you suited to this new challenge?

In this situation the project manager is a maverick within the organization. The degree of autonomy may be much greater than normal. The project manager may have to deal with very senior staff within the organization and outside. Large sums of money may be dealt with, and the traditional lines of authority completely bypassed. Despite all these problems, the plucky project manager has to deal with, and extract contributions from, all the other people within the organization whose roles have not changed at all and who may feel more than a little miffed at not getting the star role of project manager. One positive move would be to try to make sure that people do jobs that suit their personality.

The classification of people

Here is our first classification of people: project people and non-project people.

Let's take a look at project management as opposed to management in general and identify some differences that affect the make-up of people in those two environments as shown in the following table:

Project management	Management
Tasks have clear objectives.	Tasks have no clear objectives.
Tasks have an end point.	Tasks are continuous.
There is little repetition.	There is much repetition.
Role often does not fit neatly into the company structure.	Role is well established.

And here are some characteristics of people that tend to suit the two types of role described above:

Competitive	Not very competitive
Quick thinkers	Careful thinkers
Anticipate what people will say and interrupt	Good listeners
Impatient	Patient
Happy to do many things at once	Prefer to do one thing at a time

Consider their own view of themselves important	Consider other people's views of themselves important
Hard driving	Easy going
Ambitious	Satisfied with their job

Do you feel that one of those two approaches is the right one and that the other represents a weak person? I don't. It's your own views that give you your opinion. Look at your colleagues and give some thought to their attitudes to life. Are they in the right roles, or do you have some triangular pegs in some rhomboidal holes?

Is it a plane? Is it a bird? It certainly isn't a superproj. There is no superproj. There are many ways of looking at people (right way up, upside down, through a window), and there has been much work done on classification.

The fundamental idea is that there is no such thing as the perfect project manager. Such a superperson does not exist. And even if such a person did exist, your competitors would be after your superproj. You could not rely on a superperson to run your projects. The gap that would be left after his or her departure to another firm, overseas, or under a bus would be catastrophic. What you have is a superteam – a team made up of varying people with different interests, different talents, and different objectives in life. The superteam can run projects and can survive the loss of a member. It can achieve anything quickly, efficiently, and professionally.

The problem is not simply finding the right people. It is matching the roles to the people you have. A prime problem with management is having people perform, or try to perform, tasks that do not suit their nature. It is a major cause of unhappiness and discord in teams. You do not put your slim, lightweight rower on the oars and a huge hulk as coxswain. You do your best to match people to functions, and, if you cannot, you recognize that team members are doing something that sits uncomfortably on their shoulders. One of the things we can do is look at the varying types of people in order to get a clear view of the members of our teams. We shall briefly look at various highly reputed ways of personality classification.

So let's take a look at a means of classifying people in terms of their work preferences. This classification system was developed by Drs Charles Margerison and Dick McCann in Australia. Information on the topic is available from MCB University Press, Toller Lane, Bradford, West Yorkshire. It has been adapted here to the project management environment.

We are not going to try to find a perfect person. We are not even going to try to slot each person neatly into a class. If it were that easy, management would be a doddle. Everyone exhibits at least a little of every characteristic at some time. People tend to be strong in those features in one quadrant of the circle, weak in another. The topics are so arranged that like talents are grouped together.

What we can do is identify some characteristics that people in management teams exhibit strongly or weakly in order that you can evaluate their profile. Try to understand which attributes they exhibit strongly, and which they show very little of. Find their strong points. Then you can compare those characteristics with the kind of function they are trying to perform.

When classifying people in this way, do remember that people change. Age plays a factor in a person's make-up, so that, as he or she grows older, his or her profile may change. People tend, for example, to prefer technical functions early in their career and more managerial functions later on.

Let's take a look at the characteristics in a little more detail. Whom can you recognize?

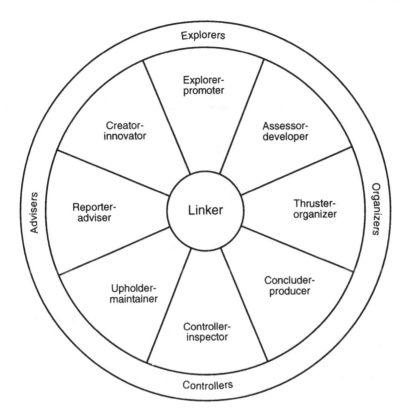

Reporter-advisers

These people are patient and persevering. They tend not to impose their views on others or lead discussions, but they do dig up information that lets others discuss matters knowledgably. Such people are thorough and find it hard to meet deadlines if meeting the deadline means doing an incomplete job. They like to get information together and feed others with it, but only if the information is complete and enough time has been available to get the information properly assimilated and ordered. They are walking encyclopedias. Sometimes they have a great deal of information, or know where to get it, but they do not volunteer that knowledge. They often like to be asked before they produce the right material. Such people are strong in monitoring and information dissemination.

Creator-innovators

These people come up with lots of new ideas. They take an active role in meetings, from time to time sparking off creative thought. Their

A walking encyclopedia.

ideas vary in practicality, and many ideas never get past the first hurdle but some good ones do. People with problems search out creator-innovators when they have a problem to deal with, and they bounce the ideas around together.

Such people are not keen on deadlines or routine; they have their heads in the clouds. Such people are not good timekeepers and are hard to manage, as their contribution seems small and perhaps erratic. They often seem not to work as hard as the rest of the team and therefore may be resented. Such people are strong at planning and at finding clever ways of doing things, and they enjoy the challenge of a new project. They quickly become bored with a maintenance operation and often with the day-to-day running of a project. Living for the future they are good at new ideas and concepts but may spend unrealistic amounts of time on topics that interest them personally.

Creator-innovators may ignore a team decision if they feel it was illogical. They may not be strong as members of a team and generally find it hard to be good managers.

Explorer-promoters

Close to the last group, explorer-promoters are not quite as good at thinking up new ideas but are better at developing the ideas of their team. They tend to be outgoing and keen to make changes, seizing on new ideas and giving them life. Explorer-promoters see the big picture and how new ideas fit into it. They are good at making people enthusiastic about some new project or idea but tend to get bored after

a little while. These people often react well to the changing challenge of project management where new projects bring new problems frequently. They are good at team motivation, like discussing new ideas with others, and are happy to listen to other points of view. They are happy to push an idea forward even if they are not always the best people to manage the change.

They are informal and unstructured. They are happy to change their ideas and systems if new information comes to light that justifies a change.

Assessor-developers

These people are good at evaluating new ideas and developing them. They are inquisitive people who like to get projects organized. They tend to be outgoing, able to communicate, and resourceful. These people are good planners and are not blinded by their own ideas. Being fairly logical, they take ideas and evaluate them, often very thoroughly. They are capable of getting together the right people and digging deep, without prejudice, to evaluate an idea. They are good at setting goals and targets. They are analytical types who enjoy having an idea to test out, or prototypes to develop. Perhaps their ideal role is in comparing and analysing a range of ideas to select the best one.

They tend to lose interest when an idea has been adopted and developed and is up and running. They are better at building a plant than running one.

Thruster-organizers

These people are good at taking someone else's design and bringing it to reality. They should be strong project managers. They are not ideal on the design side, but are better at taking a design and making it happen. They work best within established guidelines. They are good at making things happen within set procedures. Thruster-organizers establish procedures and expect others to work within them also. They are neat and tidy people who run a tidy ship. They can be very aggressive, refusing to take no for an answer if doing so means that their project will be kept on time.

They are good at keeping records or at least ensuring that records are kept. Their analytical minds lead them to set up systems to monitor the project's progress; they will also find ways to measure the project's health.

Thruster-organizers put people's needs second to what they see as the key objective – the project itself. They may seem impatient in wanting to get on with the job. Thruster-organizers make things happen even if this can be uncomfortable for others. Such people would probably make good generals. Their lack of patience may, from time to time, cause them to leap before they look. They may take a decision on the basis that any decision is better than none, when it might have been better to spend a little more time gathering information.

Concluder-producers

These people are down-to-earth, practical people who like to have everything in order. They do not like surprises but do like to know what to expect, both at work and in their private lives. They do not like planning, preferring to have plans well laid out by others. The idea of spending time looking at alternative strategies is alien to them. Meticulousness is a common characteristic of this group, making them good office managers. They set up procedures and do things on time but prefer to deal with a small number of tasks at one time. Concluder-producers may get upset if they have to leave one job incomplete and move on to another more urgent one. If this is the case, they will blame another team member for causing the problem.

These people value efficiency and effectiveness and do not get bored even when dealing with repetitive tasks. More creative people see concluder-producers as perfectionists obsessed with detail. They do not react well to rapid change but prefer executing a well-laid-out and predictable plan to guide their work.

They can be counted on to finish their work thoroughly, taking any action agreed upon to completion. They will not be late for meetings and may be regarded as punctilious. They would be very good at producing a regular progress report on a project and would take pride in ensuring that the report was complete and delivered to the right people at the right time.

Controller-inspectors

These people tend to be quiet and introspective with an eye for detail. Often happy in a role such as that of an accountant, they are happy to spend hours checking columns of figures. They are the *i* dotters and *t* crossers, and they are vital in any project. It matters less to them that the costs of the project are going over the budget than that they know how much over the budget the project is. They tend to be serious, conservative, and conscientious, bordering on inflexible.

At meetings controller-inspectors listen a great deal more than they speak, and they are unhappy with changes that affect their way of doing things. When faced with a change or a decision, controller-inspectors ask for reports to be prepared and for time to think about the change, trying to put off anything that alters the status quo. They have to be very convinced before reluctantly agreeing to change. They are good at cost or material control, keeping good, accurate records of the project's dealings.

They are worth listening to on the rare occasion that they do speak, for, while the actual topic may be one of detail, there may well be a larger question in the background. Friction often exists between explorer-promoters, who want to get on with the job, and controller-inspectors, who want to make sure that good records are being kept.

Upholder-maintainers

Such people have very firm beliefs about life inside the company as well as outside in their wider lifestyle. They are very loyal and will stick up for people. They prefer co-operation to confrontation and try hard to engineer harmony in the project team. They are usually self-effacing, quiet, and reserved. In matters of conscience they have very firm views. These people will not let people get away with running others down, demanding some proof of allegations and being defensive about others without thought for their own position.

Upholder-maintainers speak their minds in matters of principle, a habit which may well annoy the thruster-organizer, who will do almost anything to get the project finished. Upholder-maintainers only do things that fit with their views of life. They tend to be advisers rather than leaders, helping others without bothering to seek the limelight. They do not welcome change unless it is necessary.

Linkers

Linkers like to make sure that the team is working well together. Linkers take time to listen to people and discuss their work problems, helping and guiding where possible. Linkers look at the procedures and systems and get others to do so. They achieve much, often from the background. They are interested in anything that can affect the way the whole team works. Linkers often call meetings to discuss working systems and roles, and are careful to summarize everything in order that everyone understands what they have agreed on.

Linkers try not to neglect people and probably know everyone very well. The linker probably knows more than other types of manager about team members' personal lives. Linkers make sure that information is passed on properly and that a system for dissemination exists. They are co-ordinators. Linkers may have very strong team roles, but they take the time to act as linkers almost on a volunteer basis. Most people take some time to link others, but the linker makes it a special responsibility. Linking is not an easy task for a team leader, but it is better placed in the middle level of the management team, as, at that level, direct contact is easy.

Did you see yourself in that list of types? How about your boss, staff, and colleagues? Can you see people that are ideally suited to their jobs, and can you see people who are hopelessly out of tune – people who are asked to do jobs for which their characteristics are unsuitable? Once again these characteristics are neither good nor bad, they just are. They just are worth seeing, thinking about, and talking about. A better team might result, and a better team means a better project.

The bad guys

Let's face it, you do get landed with some people who exhibit negative characteristics, characteristics that are harmful to the project. Here are some people to watch out for.

Yesterday's manager

Some people live in the past and spend the whole day talking about previous projects. Somehow their previous projects were all much better than the current one, which, in their view, is not going very well at all. According to them, they did this project and that project apparently all on their own. All those projects went very well. This one isn't. They do not talk about teams they are proud to have been members of; they talk about projects they did. They tend to be senior in years but not in the management tree.

Puncturers

Some people are very aggressive and take pleasure in running others down. They criticize and never have a kind word to say. Perhaps they believe that the only way they can become superior is by pushing everyone else down. They are deflators. Leave them flat.

Outsiders

Outsiders never seem to be part of the team. They give the impression that they deserve to be paid simply for being at work, as if actually doing something positive is not part of their job. They are not proactive; the only thing they ever start is their car.

They never say much at meetings but may do a lot of whispering in corners after the meetings are over. They do not put much into the team and, for that matter, do not take much out. They do spread dissension among the team. In the few conversations they do have, they question all the wrong things: Should we be doing this project? What are we going to do after this project? Leave them outside.

Status man

This is the type who insists on a company car parking space emblazoned with their initials made up like a number plate and firmly fixed to the wall. They always come to work in a suit as it feels more important. If these persons were anything like as important as they think they are, they would be a great deal more restrained. They boast and seek status. They find excuses to spend time with really important people such as the project manager, project director, managing director, or client. They can tittle-tattle all day, spreading bad feeling, and then tell everyone how they met the PM, PD, or MD this morning. They drop names. Drop them.

Grasshopper

This type constantly changes the topic. They flit from topic to topic in a constant effort to avoid reaching any conclusion on any subject whatsoever. They can have a bad effect at meetings where their topic changes can interfere with the main business of the day. Take a firm hand and keep the discussion on course. Don't be distracted.

Boss man

This type wants to be the boss. They may be in charge only of drawings or materials, but they have a ready opinion on everything. Trivial matters such as global warming, the project's budget, and the political implications of the price of oil all have, according to the boss man, simple remedies. These simple remedies have a wonderful feeling of workability about them, but are never tested as the boss man never asks for or gets a position of power.

Such people should have been journalists or art critics. One thing they do not believe is 'Those who can, do; those who cannot, talk about it.'

Here is another way of provoking a little thought about project managers. What should a good project manager be?

If you are going to be a good project manager, you must be:

intelligent;
proactive;
self-assured;
a helicopter communicator;
a helicopter thinker;
persuasive.

Intelligence is a key factor and a fairly obvious one. Perhaps streetwise would be a better word, as you are going to have to keep your wits about you to get on in the wonderful world of project management.

Proactive means the ability to shape the future – to go out and make things happen. Managers can be reactive – i.e. they can wait until things go wrong and then react to them. You must go out and make things happen – otherwise nothing will happen. Project managers have to fill a vacuum with activity.

Self-assurance is a useful characteristic, as you may well have to walk into some rather imposing office belonging to some rather important person and persuade that person to do something that helps the project along.

Not only does the project manager have to discuss the project with VIPs; there may also be a need to stir a craftsperson into action. The good project manager wants everyone on the project to be reasonably enthusiastic about the project. This means the project manager has to be able to discuss things with both labourers and managing directors Hence the term **helicopter communicator**. You must be able to move up and down the social scale, talking to everyone and anyone in order that your project moves forward.

Similarly, you must be capable of **helicopter thinking**. You need to be able to think about major matters – the funding of the project and the overall project plan – as well as about more prosaic matters – matters that might affect only one welder. Everyone will expect you to know the answer to everything. Everyone will expect you to be able to solve everything. To keep you in good shape, people will present you with questions that you should not be deciding about. Beware when asked for a decision about something; check to make sure the right people make that decision.

Regardless of their position in society, you must be able to persuade staff to perform in the best interests of the project. You need to be able to sell your ideas so well that everyone thinks that the ideas are theirs, not yours.

Decision making

Decision making is as essential as breathing. We make decisions every day; most of them we are not conscious of, or are barely conscious of, making. We decide when to get up, what to wear, what to eat, and when to leave for work. We spend a lot of time thinking over less routine decisions, for example, what to do with our careers, what to do about relationships, and what to do about specific problems. Decisions are like gates on a highway; we make no progress without negotiating them.

Project management teams face a wider variety of decisions – they are constantly deciding how to do something; how some operation should be achieved and with what equipment. Every decision should be made after due consideration of the effects on the project in terms of cost, quality, and time, Most decisions affect all of these three factors, and it is partly for this reason that the project team is structured in the way it is.

There are two extremes to our decision-making ability – we can use intuition (associated with the less logical, right side of the brain), or we can use reason (associated with the more logical, left side of the brain). The diagram on p. 164 shows how these two approaches differ. Intuition refers to a hunch, what feels right by the seat of your pants. Reason

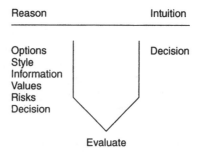

means following some logical progress to reach a decision. Groups of people working together tend to use reason.

You can use both approaches to a problem – we should be conscious of what we feel about a decision as well as able to use reason. The degree of each may well be dictated by the decision to be made. There may be actual information to help a reasoned approach, or there may be very little experience or few data at all.

Decision making can also be classified into two types. It helps sometimes to think about decisions in these terms to see what kind of process is most suitable.

Types of decision

Long-term	Short-term
Strategic	Tactical
High-risk	Low-risk
Qualitative	Quantitative
Intuitive	Logical

If you do decide that a formal decision-making process is required, the following steps may help you approach the decision in a structured manner.

The decision-making process:

Identify problem
Establish criteria
Select alternative
Gather information
Evaluate alternatives
Select compromise

IMPLEMENT DECISION

Motivation

Motivated people are productive and enjoy their work. They achieve satisfaction from achieving or striving towards the group objectives. There are many factors that can motivate people, but you, the manager, are not one of them. People are naturally motivated, and all you can do is to understand their motivations and try to fit in with them. There are some helpful theories about motivation which we can only touch on here.

Abraham H. Maslow wrote a book called *Motivation and Personality*, which was published by Harper and Row in 1954. Maslow thought about what motivated people and came up with what he called the **hierarchy of human needs**, of which there are five. By recognizing these levels, you can try to help people be motivated.

Starting with the lowest, we have the human being's physiological needs. We need to eat, drink, and be reasonably dry and warm. Most people in the West have achieved this level. Many others have not. After these needs are satisfied, we need safety and security. Assuming we have food, we worry about being eaten by a sabre-toothed tiger, mugged, or sacked. Now, many people feel that a lack of security is highly motivating, but we shall come to that later.

Assuming we feel reasonably secure, we worry about love and belonging; our social needs emerge. We wish to be part of a group, a family, or a team, and we react to protect and defend our groups. These are the sorts of things that project management teams need to worry about. Our ego and self-esteem come next up in the hierarchy. We like to feel good about ourselves, and we want others to feel good about us. The fifth level is self-actualization.

Frederick Herzberg wrote in his *The Motivation of Work* (John Wiley, 1959) that there were dissatisfiers and satisfiers in the workplace. He asked 200 engineers what factors in their jobs contributed to job satisfaction and what factors produced dissatisfaction.

He found that the reasons for dissatisfaction were hygiene factors, reasoning that a lack of hygiene produces illness, but a presence of hygiene does not produce health. Hence, a lack of job hygiene avoids dissatisfaction, but it does not produce motivation. There should be positive motivators as well as a lack of bad hygiene in an effective management environment. The trick is to recognize which factors are hygienic and which are not; which factors are dissatisfiers and which are satisfiers. The following table gives the essence of the book that Herzberg wrote:

Dissatisfiers	Satisfiers
Company policy and administration	Achievement
Supervision	Recognition
Working conditions	The work itself
Interpersonal relationships	Responsibility
Status and money	Advancement

Surprised by the last one on the left? Status and money are dissatisfiers. Their absence causes problems; you quickly get used to your current level of status and income. It is only the prospect of getting more money and status that excites and motivates.

I am a follower of Edward de Bono and his school of lateral thinking. This is not a management theory but a way of thinking, and I recommend that you buy one of his books. At least you will find de Bono's books in your local W.H.Smith. He talks about the way we think and the way we approach problems. He says that dealing with a problem is like finding your way through a town. You make assumptions (take turnings), and each assumption takes you into an area of thought (down a side-road). If you get lost and cannot find a solution, you should go back along your route and see what assumptions you made (which turnings you missed) without ever realizing that you made them.

One of his best examples is that of the watchdog. Apparently, a man bought a watchdog, but when the dog was on duty it never barked, and it was therefore ineffective. What could he do – replace the dog, fix a device on the dog so that when it moved it would wake him up, or fit an electronic surveillance system? De Bono advocates the lateral solution, getting people to think clearly about the objective and checking their assumptions. The man put up a chilling sign: 'Beware of the silent dog.'

There are three other management theories worth mentioning. One theory likens the company to a tree. The board or shareholders are the roots, the trunk represents the senior management, the branches represent the various levels of middle management, and the leaves represent the field operatives. Alternatively, in the wheel theory, the hub represents the directorate, the spokes the middle management, and the rim the various layers of operational management. Finally, in the mushroom theory, the management keep you in the dark and shovel manure on you twice a day.

If you want to delve deeper into classical management theories, visit the library; there are many more than these two books mentioned above.

There are some very common-sense steps to getting the best out of people by giving them the opportunity to do a lot for the project. When we talk about people we are talking about you and me and your Cousin Neville. People in companies are real. Make people feel important in themselves – they must not feel like replaceable equipment. They like to know what is going on and that they can contribute.

They need to feel that their contribution has a direct effect on the company. This is very hard in those organizations which are very large or which have no means of measuring their own success. Junior workers in a very large public body have no way of measuring the effect of their contribution. People like to feel that as the company grows, they also grow. If the company does well, people like to feel that they also benefit. On the other hand, I am not sure that people are willing to suffer if the company is not doing well.

People like to feel that they are part of a team, not just underlings. As far as possible, senior people should make sure that they know everyone and what everybody does. People like to feel proud of their organization. They like to be able to talk about the work they do at dinner parties and in the pub.

Programme management

Many organizations find it necessary to manage a number of overlapping projects at any one time and often these are expected to deliver change to the host organization. This environment is notoriously difficult to cope with and is a significant topic in its own right but one that this chapter attempts to outline.

Introducing programme management

We can neatly divide the world of project management into two groups:

A: organizations that perform projects because they are paid to do so

B: organizations that perform projects to improve themselves.

If this causes you to furrow your brows you have probably spent your working life in one of these two camps and had not realized that there was another group at all.

Type A organizations, for example a building contractor, a software developer, a ship builder and a pipe fitting company, all exist to run projects for their clients. They win work through some form of bidding or negotiation, do the work and get paid for their troubles. Success will be measured in terms of the profit the contracting organization mades out of the project. We can describe these as **external** projects as the outputs are delivered to another organization. Many people within the organization are employed to manage projects on a full-time basis.

Type B organizations, including a bank, a government department, a supermarket chain and a hospital, will identify and initiate projects and spend money to achieve them. The organization will not be paid for the work but will expect to benefit in other ways. These benefits might include increased efficiency, bigger profits, improved service levels or reduced costs. Success will be measured in terms of investment and benefit. It is very possible that such a type B organization will use a number of type A organizations to deliver these projects. We can use the term **internal** projects as they are managed within the host organization with the aim of improving the way the organization operates. Very few (or no) people are employed to run projects and everyone has a normal job to do, as well as helping out with those pesky projects.

Programme management is primarily concerned with type B organizations where internal projects are managed on a part-time basis by project managers in their spare time trying hard to improve the way their organization operates.

Organizations that run internal projects without programme management are easily recognized by these symptoms:

1. Most people have only the vaguest idea of what projects are going on, what they are designed to achieve and what impact they will have on the organization.
2. Odd projects spring up all over the place in an unstructured form. Some absorb loads of time and effort, some fade away and some do both. If projects were flora, there would be few roses amongst the weeds.
3. Projects are run in an inconsistent way without controls or systems.
4. It is virtually impossible to identify clearly any part of the organization that has improved due to effective projects.
5. Projects get the go-ahead because they have the backing of an important and influential manager, not because they are vital to the organization's strategy.
6. Some projects seem to be doing the same thing – the 'wheel is being invented' many times over.
7. It is easy to have a quiet life pretending to work on some projects but actually doing the crossword and writing a blog.
8. No one can confidently say that all projects can be delivered with the resources available.

Does any of this sound familiar?

Programme management tries to address these problems. I am not going to claim that organizations that run their projects with programme management have none of these problems; there is no magic wand. But there are organizations that run their workload brilliantly and just can't understand the problem.

My preferred definition of programme management is:

Programme management is the management of change.

The UK government has published a book on this topic called *Managing Successful Programmes* (The Stationery Office, 2003) and it contains this slightly more wordy definition:

Programme management is the co-ordinated management of a portfolio of projects that change organizations to achieve benefits that are of strategic importance.

Programme management breaks down in simple terms into three facets:

1. Project Portfolio Management (PPM)
2. An environment within which projects can be successfully managed
3. Benefit management – extracting and delivering the benefits from those projects

Let's outline each of these three, remembering at all times that we are concerned with delivering change within our own organization.

1. Project Portfolio Management (PPM)

The lowest hanging fruit in most organizations lies in the way projects get started. At worst, projects just spring into life as a knee-jerk reaction or as a result of a bright idea or a political imperative; this is especially common within government organizations.

Slightly better projects are 'justified' by someone estimating that the cost of the project will be returned many times over. But this process of justification answers this inadequate question: is this a good project to do?

PPM asks a better question: what is the best possible group of projects that we are able to deliver and that will most help our organization to achieve its strategy?

PPM is not a simple process to establish but it is cheap in relation to the fantastic waste of money that goes on in many organizations.

One American study showed that only one in four projects undertaken by the IT departments of some major banks delivered a benefit. What were the other three for? But wait – here's worse: many organizations would be delighted to hit one in four!

PPM suggests a process where all ideas for projects are analyzed in a consistent and thorough manner before any work starts. Each **contender** project is put under a microscope to establish:

- Mandate – a written statement of the purpose
- An owner and/or sponsor
- A list of benefits:

  - For financial benefits

    - No change cost or income levels over time
    - Post change cost or income levels over time

  - Non-financial benefits – measures of strategic alignment through key performance indicators

- Risk estimates (schedule risk, cost risk, benefit delivery risk)
- Financial investment required
- Resource requirement
- Stakeholders – those involved with and impacted by the proposed project
- Other projects impacted by this proposal

Often this involves spending some money on a **discovery** project designed to investigate the idea more fully. Ideally this work is done by an independent expert group, not the over-enthusiastic person who had the idea in the first place.

Ideally the groups that will be affected support the proposal by agreeing to the benefit proposals (see the later section on benefit management).

Once the project has been defined it goes before a programme board where projects and programmes are selected and prioritized.

The programme board have a clear understanding of the organization's strategy and seek ideas for programmes and projects that will help them to deliver it.

Only with the approval of the programme board does the project get the green light and proceed. The programme board will publish a project schedule showing all the live projects (except perhaps some secret ones) and make this widely available. In such organizations no work is done on unapproved projects.

PPM is not a one-off process but an on-going cycle. In many organizations the programme board meets quarterly to examine all live and contender projects.

The programme board may cancel, re-define or delay live projects and approve new ones. They understand the current organizational strategy and the environment in which the organization operates.

2. An environment within which projects can be successfully managed

Programme management people do not micro-manage, they leave the projects to the project managers. The programme management team have the responsibility to ensure that every project is clearly defined and managed. This will imply that every project has the following elements all neatly lined up:

- A clear objective in terms of deliverables and benefits
- A clear timescale
- An established project manager who is given time to manage in addition to any other workload
- Access to the resources it requires

- A known and understood method
- Access to a project office or programme support office where expertise on project management matters is available
- A requirement to regularly report to the programme board
- A process for escalating issues and risks
- Defined quality standards
- The tools and knowledge that people on the project need.

In an organization where the programme management team do their job well, running projects is simple. Project managers should feel valued and supported and understand what they are doing and why.

> Does this sound as much like your organization as The Clash sounds like Tchaikovsky? What are you going to do about it?

3. Benefit management

Parts one and two above describe an organization that carefully and logically selects its change programmes and then provides an environment where projects and programmes can be successfully managed.

The third leg to programme management is the harvest. It is the bit your organization wants in the first place: the benefits.

Neither a programme nor a project can deliver a benefit. Get that idea out of your head – I don't even know where it came from. I pity you if you are a project or programme manager and have responsibility to deliver benefits. It is time to run for the door.

At the very, very best, the relationship between project, programme and organization follows the diagram below. I call this the value path.

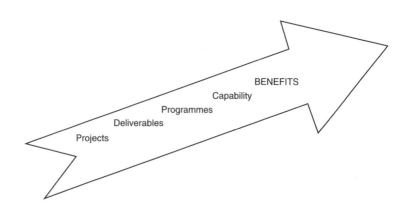

The value path suggests that projects create deliverables. The deliverables are collected together by the programme management team and thus create a capability. This capability is then handed over to the organization and, if (and that is a very big if), the organization uses the capability, the benefits are delivered.

Let's take a hospital as an example:

This hospital is trying to reduce waiting times and has a clear idea of how long patients have to wait for different types of operations. The programme board decides, after careful consideration, to build a new operating theatre. This programme is broken into a number of projects, each aligned to the functions of the organization including:

Component project	Assigned to:
Architectural design	Architects' or estates' department
Construction	In-house construction management team employing a construction company
Specialist environmental services	In-house environmental services team employing an engineering contractor
Operating theatre equipment	In-house equipment team and suppliers
IT support	IT team
Staffing	HR team to recruit staff
Commissioning	Temporary project team to establish protocols, stock up consumables and other disposable equipment and goods

Once all these projects are finalized and complete the new operating theatre (the capability) can be handed over from the programme manager to the operating theatre management team. The operating theatre management team clean the place up, make sure everything and everyone is ready and start accepting patients.

The programme management team stand back in admiration but scrutinize the size of the waiting list over the next months and years. The programme management team support the operating theatre management team and help them to measure the size of the waiting lists but in the end, the operating theatre management team have the responsibility of delivering the benefit of shorter waiting times. The programme management team were responsible for delivering the right operating theatre along with the IT support and equipment that the surgeons will need.

Therefore project and programme managers cannot deliver benefits. They can only deliver the capability and monitor the effect of the change.

What are benefits?

Benefits are measures of improvement within an organization. Some are financial, some are not. A project can make many benefits possible. A single benefit may depend on many projects.

Benefits should be agreed upon between the department or departments that will be affected and the programme manager who will deliver the capability to change.

Financial benefits include reduced running costs and increased income.

Here are some simple examples:

- The accounts department might agree that if an IT project delivers the new payroll system, their running costs will be reduced by 5% over the next five years.
- The sales and marketing team might agree that if the project delivers a proposed new product in August, they will increase sales in the pre-Christmas period by 10% over last year.

Often benefits are non-financial.

- The local authority cleansing department agree that, given a new fleet of trucks, they will reduce complaints from irate householders by 7%.
- The safety officer agrees that as a result of the proposed safely training project, accidents and absence caused by accidents will reduce to 3%.
- Everyone in a phone manufacturing company recognizes that whilst the new video phone will raise sales in the high street marketplace, it will cause a reduction in sales of the outdated model through on-line shopping.
- The director of retail operations agrees that the new high street store card will increase customer retention by 2.5%.
- The factory manager agrees that the new testing process will reduce wastage by 5%.

Running Cost Predictions

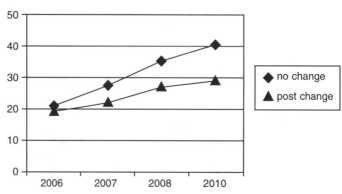

Financial benefits are best understood by comparing what is expected to happen to costs or income if nothing changes and comparing this with what is expected to happen to cost or income after the programme.

In this example, the running costs of a part of the organization are expected to rise over the period 2007 to 2010. If the proposed automation programme ends successfully, the cost increases will be significantly less.

A few points to note:

- Benefits do not relate to the projects or the programmes that make them possible. Benefits relate to the purpose of your organization.
- Benefits are measured in terms of improvement to the way you work. They measure some form of change, hopefully an improvement, in the way your organization operates.
- Finishing a project on time is not a benefit.

We tend to focus on project schedules and end dates, mostly because they are easy to measure. It is much better to focus on the expected benefits as this is the true objective of such projects. There are projects where time is important (fireworks are much less useful on the shelves on 6th November) but often benefits will go on for years to come.

Non-measurable benefits are the weakest type. Making qualitative statements like 'we will have a better public image', 'staff turnover will reduce' and 'the workspace will be safer' are nice, sometimes unavoidable, but hard to prove and disprove.

> The value of the benefits achieved will last long after the sweetness of finishing on time has been forgotten.

Finally here is a quotation from the 'Improving Programme and Project Delivery (IPPD)' report from The UK Prime Minister's Office of Public Service Reform (OPSR) (2003):

> That's why the implementation of this report is so important. I am pleased that one of its key recommendations – establishing Programme and Project Management 'centres of excellence' within departments – has already been endorsed by the Cabinet.

Issue and risk management

Within these pages the author's views on risk, and, for all that, personal views, are outlined. The tools available to the project manager wishing to estimate and manage the unmanageable are described and their use prescribed.

The career critical question is not: Should I be doing some risk analysis? The career critical question is: How much risk analysis should I be doing?

I can load you down with statistics that support the idea that risk is worth a look. I could quote *Computer Weekly* which said that 'Up to 50 UK companies recently lost as much as $1bn between them as a result of abandoning or replacing projects'.

I could mention another survey of IT management consultancies which showed that only 30% applied any form of risk analysis and yet 90% of projects went over budget and 98% had changed specifications, usually to a lower spec.

This is all sensational stuff but a more careful reading tells me that these quotes say very little. No one doubts that there are problems within the IT industry but these statements don't tell us how big the problem is and nothing tells me that risk analysis is going to be the magic wand that brings sweetness and light. For all I get from reading these quotations, risk management might make things worse.

Despite these doubts I still believe risk analysis to be a very useful tool.

I hope that by now you have noticed that I don't lead you to do things simply because they are better project management, or even because they might result in a better managed project.

Risk analysis is 'career critical' and that's what I call important. Risk analysis can make the difference between a job and no job.

Here's the official line:

- In any project, ideally at project definition stage, the team should invest some time in evaluating risks.
- There is great value in understanding what investment and work might be at risk if the project is abandoned and what delays and overspends might result if potential, identified events come to pass.

Here's why risk could save you from the jobseeker's allowance.

A risk identified is a risk shared. If you and the rest of the project team identify and discuss and collectively accept some risks, the team, the client and the programme management will all go ahead with the project with their eyes wide open – with the risks understood and acknowledged.

If you do not detect a serious risk or don't mention a big one, you are taking that risk on your own shoulders which is unfair on your project's senior management and the project sponsors. If you identify no risks you are taking the whole lot on your shoulders. And Head & Shoulders is not going to clear this stuff away.

This is career critical stuff. If you talk about risks with the powers that be and then you all agree to proceed, you are all in it together. If something goes really wrong and you can reasonably say: 'Well, the thing we all feared has come to pass and it is causing some significant problems – it was a risk we took. The contingency plans we laid will be brought into play.' This is great management speak and should aid your voyage to the great golf course on the top floor.

On the other hand if, at the start of the project, you presented a glowing case to the board and pretended that nothing could go wrong and then something big fails, they will be within their rights (if the board need to be in the right) to say: 'If we had realized that the project would fail if (substitute the most recent disaster) we would never have proceeded. You are fired.'

It is the last three words that really hurt.

I mentioned earlier that this risk analysis stuff should be done early in the project life cycle. Two reasons for this spring to mind – one is that you might be in the 'seeking approval' stage and you should get these risks on the table before the approval is given. At least the approval acknowledges the risks you have identified. Anything found later may be too late to save your hide. Also the cost of fixing problems gets bigger and bigger the further into the project you get. Early fixes are cheap fixes.

So a simple statement of the risks that can be foreseen is a useful part of a project definition. It is very easy to go overboard with this concept and list every possible risk. You can spend happy hours with your team dreaming up all sorts of dire events that could wreck your project. It is actually quite fun in a macabre sort of way – it's like imagining all the awful things that could happen on holiday – and eventually you have to start laughing. The list of potential disasters can grow and grow but misses one serious risk: as soon as the senior management see this huge list they will get the collywobbles and cancel the project all together.

If you do too big a list you will probably frighten everyone off and your project will get no backing at all. As the trapeze artist said: 'balancing is everything'.

You have to balance your wish for the project to go ahead with the need to investigate the risks that you will all face. So if you are now convinced that some risk analysis might be useful, here are a few guidelines explaining what you can do. If you are not convinced skip to the next section.

One other thing for those who have decided to read on – risk techniques are supposed to assist management, they are decision support techniques. Risk tools do not take decisions but are aids to decision making. Don't believe what some of the software vendors put in their brochures.

Risk management can be broadly separated into two classes: qualitative and quantitative. I'll take them in alphabetical order.

Qualitative risk

Qualitative risk refers to the general type of risk that can be imagined and foreseen but can only be discussed in general terms. Sure, you can put some simple numbers to some of these risks but the mathematics is going to be dead simple.

Qualitative risk, at its simplest, involves only a description of the obvious risks being undertaken as a part of the project.

Techniques do exist to help understand and locate these risks, for example the herringbone diagram (does this look like a red herring to

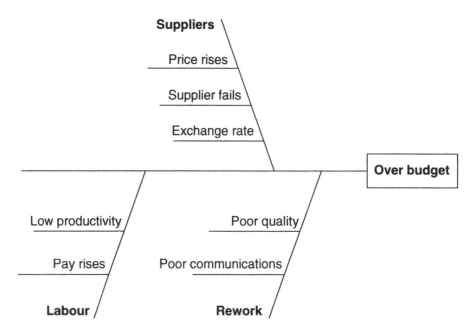

you?) where an effect relating to the failure of the project is chosen and those elements that are likely to lead to that effect are indicated.

Here the problem of going over budget is identified as a potential failure. The risk analysis group sit about and think of the causes that could send the project over budget and find three major categories. They are: labour, rework and supplies. Labour costs can be looked at in a little more detail – a pay rise might put costs up and low productivity might make everything too expensive.

I hope you've got the idea – you think of something that would be a failure and then think of things that might cause that failure to happen.

The diagram's purpose is:

- to promote useful thought;
- to provide some form of structure;
- to allow you to use impressive terms like 'herringbone diagrams'.

I believe in drawing the team together to talk about risks as a few minds will unearth things that individuals would not think of on their own. Almost any time spent thinking ahead about the project is useful. You can impress them with your knowledge of the techniques.

You can draw such diagrams for finishing late, not meeting the need, not achieving quality, not meeting the company's standards and many other failure modes.

Once risks have been identified, a simple mathematical approach can be used by giving each risk two values on an appropriate scale. This is an attempt to bring a little science into the picture. One value relates to the likelihood of the risk's manifestation and the second value relates to its impact on the project. The first is a measure of how likely the thing is to go wrong, the second is the effect it will have on the project. By multiplication of the two values you get a weighting for each risk.

Putting this information together in a table is sometimes called a **risk register**. A risk register can cover more ground in that it can discuss at length the nature of the risk, the impact and the things that can be done to prevent or reduce the impact of the risk. Here is a very simple example for a small house building project in Surbiton. The values are rated out of ten.

Risk element	Likelihood value	Impact value	Weighting of hazard
Front gate not available in chosen design	9	0.01	0.09
Earthquake	0.01	10	0.1
Heavy snow	3	3	9
Heavy rain	6	2	12

Weighting equals likelihood × impact. Some people call this derived value the **hazard** or the **hazard rating**.

There are events that would ruin the projects but are very unlikely – the earthquake for example. An earthquake would ruin the project but they are rare in Surbiton. There are people who believe that this would be a better place if there were more earthquakes in Surbiton but surely it cannot be that bad a place.

There are events that are very likely but have only a small effect, e.g. not finding the preferred gate is something that is quite likely to happen. Its effect would be quite small as an alternative could easily be rustled up.

This technique allows each risk to be listed in order of greatest hazard. Greatest weighting is a term some people use for this.

You can extend this table with a couple of extra columns if you are in the mood and, even better, put some meaningful information in those columns. How about ownership? Some organizations talk about ownership of risk – indicating the person who has responsibility for dealing with this risk. This column is an extremely good place to not have your own name.

You could also have a column mentioning secondary risk or 'knock-on effects.' An example might be that high winds over 14 knots will cause the crane to stop working, the secondary effect of which will be reduced stockpiles of heavy steel stock as without the crane, unloading will have to stop.

Such analysis is a useful part of a project definition and is a worthwhile exercise to carry out early in the project (especially when you take over someone else's project). You end up with a one or two page overview of the major risks on the project which forms a part of your project definition document. Once the project gets approved, the risks are all nicely shared and you have done your best to be a maestro project manager.

This kind of risk stuff is not always easy and it certainly is not accurate. Many types of risk are very difficult to measure in any kind of meaningful way and your subjective viewpoint will probably colour your judgement when asked to give a risk a score out of ten. Some risks are very numeric – the weather records can be analysed and accurate predictions made.

Risk management is not a static, one-off process; it is a continuing part of managing a project.

As work proceeds, your ideas about risk will change. Some risks will get more likely and/or be predicted to have greater potential impact. Others will fade away. As work gets done some risks will completely disappear. New ones will emerge.

Risks need to be managed. That does not mean that you try to eliminate them but you might take steps to watch them, reduce their likelihoods and/or their impacts. You might delegate responsibility for this management to a colleague.

All of this requires management so please add the item 'risks' to the agenda of your regular project review meeting. Spend a little time updating the risk register and discussing what, if anything, can and should be done.

Quantitative analysis: risk and critical path

If you have a critical path diagram for your project you might like to hear that some people say that a project is a high risk project if a high proportion of activities have little or no float. They say these things because critical activities, if delayed, delay the project whereas activities with float can be delayed without direct effect.

Hence if many activities are on or near to the critical path the project is likely to be a high risk project. This is much more of a problem in physical projects than in non-physical and even more of a problem in physical refurbishment type projects where the workload is very unpredictable, but in most projects it provides a useful rule of thumb.

There is more that can be done with a critical path diagram on large one-off projects and this brings us to **quantitative risk analysis**.

Quantitative risk analysis brings risk and critical path analysis together along with mathematics, probabilities and Monte Carlo techniques. Here you must start off with a critical path diagram of the project created within the bowels of a project planning software package. These packages bear the tag 'project management software' but it is just not true – they are only planning tools.

The tasks in a critical path model are jobs of work that need to be done and are described by a description and a single estimate of duration. Links show how some tasks depend on others. The critical path analysis engine which forms a part of every project planning software system churns through the model and decides when each task can and must start and finish and which ones are vital to the success of the project. It will elect some tasks to be critical which means that any delay to such a task delays the project. The length of the critical path is the overall duration of the project.

The problem with all of this is the single estimate of a task's duration. How long will it really take to design the googgleflit? When we remove the inspection hatch, will we decide simply to clean the motor (2 days) or replace it (2 months)?

In quantitative risk analysis the single estimate of duration for each task is replaced by a number of possible durations. Sometimes a range of likely durations is given to each task. You can say things like:

- The most likely duration for this task is 6 days, the optimistic duration for this task is 3 days and the pessimistic duration is 10 days – this is the original meaning of programme evaluation and review technique.
- There is a 10% probability of finishing the task in 4 weeks, a 75% probability of finishing it in 6 weeks but a 15% chance that it will take 3 months.
- The graph of probability against likely duration for a task looks like this:

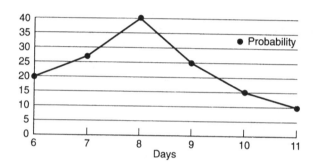

These kind of estimates can be made for each and every task in the plan.

In addition to a range of possible durations for each task, some systems permit a range of possible logical connections. Instead of the simple finish to start dependency you can select a number of different dependencies each of which has a probability and each of which takes you off down a different path through the plan. You'll read of an example of this in the case study that follows at the end of this chapter.

Now that each task has got a range of durations set against it and we have worked out the possible logical routes we can give Monte Carlo a look in. Monte Carlo immediately brings up certain images in my mind none of which I am ashamed of. They include handsome men in tuxedos, beautiful women in evening gowns, Grace Kelly, casinos and velour, suave and debonair people in expensive surroundings occasionally shooting themselves on the balcony.

This in turn brings up images of the local cinema and popcorn because that's as near to Monte Carlo as I get.

It all sounds unbearably romantic and upper class and people fall over themselves to get in on this act. When you do find out about Monte Carlo techniques in the world of risk analysis it is as disappointing as saving the last cream cake for later and then finding it has gone bad.

The Monte Carlo technique means random numbers. People make a living creating elegant, ingenious and dramatic methods of picking random numbers and the name comes from the very random numbers generated by the roulette wheel and ball bearing so long associated with the South of France. They could have called it the Las Vegas technique or even the roulette but this would have sounded much less romantic.

To the average project management person the technique used for selecting the numbers counts less than the price of cheese. It is a shame, hopelessly dull and I hope you're not too disappointed. It is true. My level of interest in how the random numbers are chosen couldn't be lower – I have complete faith in the clever software people who spend their time dreaming up algorithms for things like that.

So to carry out a quantitative risk analysis you take the critical path model, add a variety of durations and logical paths to each task and feed the whole lot through a risk analysis software package. The project plan with its many tasks, each of which has a range of durations, is analysed many times – each time using a randomly selected duration for each task within the stated parameters.

Each time it analyses the plan it derives an end date using a selection of possible durations and logic routes. Round and round it goes hundreds or maybe thousands of times, each time coming up with an estimate of overall duration. Each go at analysing the plan is called an iteration and the technique is called an iterative technique.

Some analysis runs will give long overall durations and some will come out much shorter. Eventually the most likely overall project durations emerge. On some software packages a graph showing end dates against likelihood of achievement is produced.

The whole process is often called **risk analysis** and tends to be used on major projects where risk plays a very significant role. Remember, it is only applicable when a critical path diagram is in use on a suitable software package.

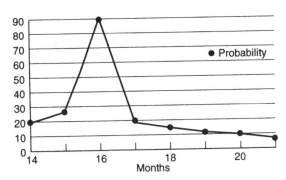

The pretty graph shows that this project is 90% likely to finish in 16 months but could run on to 20 months and there is a 10% probability of that happening.

You could then use this graph to predict not just a simple end date for the project but a range of end dates each with a probability. Then you can think about the cost of any steps that might improve probabilities with a sound basis to judge their value. What would happen if the bungalow in Surbiton was built inside a huge tent? The risk associated with bad weather would almost disappear and the effect that would have on duration could be calculated. You could then decide if the tent is worth building. You could also think about the effects of building in winter or summer.

There is something else that comes out of the risk analysis engine which is kind of fun. As the process involves selecting durations for each task and then analysing the plan, each analysis may result in a different critical path. If you do 1000 runs you will have a good choice of critical paths. Some tasks will insist on popping up on nearly every critical path and some will appear very rarely. You can say with some confidence which tasks are most frequently critical and therefore deserve maximum attention.

Quantitative risk analysis is mostly about time. It concentrates on probabilities and durations. You could deduce the budget implications of each duration and find out how costs compare with duration and with probability.

Before you dive into the deep end of the pool marked quantitative risk analysis, think about costs. It is going to cost a lot of money to get set up and do the first analysis run. And when the project has been underway for a few months and the actual progress bears only a passing semblance to the original plan, are you going to rerun the risk analysis studies? If you do you will have a regular bill to pay.

While on the thorny question of money some people will spend happy hours carrying out exercises similar to that described above but using money in place of time. You can see the exercise is very similar. You have a budget made up of elements and at first you have your best guess at an amount of money for that element. You can then attach formulae indicating probabilities and their associated costs before performing the famous Monte Carlo analysis on the budget. In such a system, the software adds up the costs many times and gets a distribution of cost against probability.

These two risk analysis techniques are not simple choices. On some projects the team might use qualitative techniques early in the project's life cycle and then apply quantitative techniques later on during the detailed planning phase.

Software packages that do this kind of risk analysis include Predict! from Risk Decisions and Pertmaster. Some people buy a risk analysis software package and others hire a risk management consultant who comes armed with expertise and software.

Some people, by the way, talk about two types of risk. They talk about implementational risk and benefit risk.

Implementational risks are the risks you might encounter within the project itself. For example will the deliverable be delivered on time, will its cost more than approximately coincide with budget?

The benefit risk is concerned with the thing that remains after the project is over. Did the end result of the project – the deliverable – actually achieve its goals, did it meet its objectives? There have been magnificently run projects which came out bang on time, bang on budget and delivered exactly what was asked for but completely failed to help the company move forward.

Let's say the objective of a project was to reduce transportation costs by 5% and the selected means to achieve this was a new warehouse in Milton Keynes. No matter how beautiful the warehouse, no matter how well the project was run, if the warehouse does not have the desired effect on transportation costs the project will have failed.

In these terms it is not enough to run a tightly controlled project, you have to keep your eyes on the benefits as well. These can conflict with each other. A software development project might be held up as there is a problem with the quality of the software. Perhaps some part of the program is not working properly and a specific report is not yet reliable. This is an implementational problem and it causes a delay to delivery. The marketing people might have a window in which to maximize the launch of this new service and would happily do without the specific report as long as they can launch the new product on time. The implementational risks are in conflict with the benefit risks.

That's when it is a good time to separate the two kinds of risk in your mind.

Case study: the British Library

This case study is taken from *Project Manager Today* and I am grateful to that magazine and the helpful people at the Library and at Eurolog Limited, the Teddington-based risk experts, for their help with this case study which was written when the library was under construction.

What was the last great public building before the new British Library?

Britain is well supplied with great public buildings. They are structures organized and built by the government of the day and for the public good. Westminster Palace, St Paul's, The V & A, British Museum and The Royal Opera House are all well-known examples. The New British Library on Euston Road and next door to St Pancras will one day surely rate among the great.

PSA Projects are managing the new library project on behalf of the sponsor, The Department of National Heritage, and the client, the British Library.

What we especially wanted to look at on this project was the team's use of quantitative risk analysis techniques and so your faithful reporter trotted along Euston Road one wet afternoon to meet James Macrae of Eurolog – the project's risk analyst.

James's background is in civil engineering. With Taylor Woodrow International he worked in Dubai, Nigeria and Trinidad in a project coordination role. He recently managed to find time to get an MBA at City University Business School before joining Eurolog in 1988. Eurolog and the MBA took him out of construction into the worlds of defence, telecommunications and IT so he welcomes the library as an opportunity to return to construction.

The need for risk management on the project stems from the complex interfaces between the construction, commissioning, handover and occupation phases of the building. The library team have a great deal to do once the building is complete – the new computer book retrieval system has to be installed, tested and got working as does the book delivery system. This may not sound too hard until you remember that there will be 11 million books on the 300 km of shelves in the four deep basements underground. Let me take you forward in time on a short guided tour of the new building.

When the building is in use you will enter through the main hall from the piazza area on Euston Road having already walked above the deep basements in which the books will be stored. The main hall is very grand with marble pillars reaching from the ground floor entrance area to the ceiling many floors above. Balconies along the rear wall of the hall connect the two main sections of the library to your left and right.

Proffering your library ID card you will turn left for the Humanities section or right for the Sciences. In either case it is likely that you will seat yourself in a reading room at one of the many public computer terminals and search the computerized database for the book you want to consult. Once you have made your selection you will press a 'Please get it for me' button which causes a slip of paper to be printed out some distance below you in the depths of the basements at a librarian station some-where near where the required tome is stored. One of the librarians will respond to your call taking the book from its place on the shelf and sending it off on a journey using automatic paternoster lifts and con-veyors running in hidden ducts and tunnels throughout the building. Within 20 minutes your book should appear at the reading room counter nearest you and a red light on your terminal will let you know that you can stroll over to the desk and get your book.

This is a significant improvement on the current 24 hour wait that you can expect while the book is delivered from one of the many book repositories that will be replaced by the new library.

In addition to getting all the book handling systems installed and up and running, which of course means new staff and staff training, the building has a very sophisticated air conditioning system and fire protection system which will require some looking after. The air conditioning will maintain the building's environment and is linked up directly to some of the bookcases where it will help to maintain the more precious volumes (the actual Magna Carta and Gutenberg Bible to name an unreplaceable two) in peak condition.

So the library and the construction team have a job on their hands getting the book handling systems and building services into place. Add to these commissioning activities the normal administration systems that need to be brought into life – security, accounts, personnel, canteens, toilets, cleaning, etc. – and add in the phased handover of the building as various sections get finished off by the building team and you have a king-sized planning problem.

It is planned to hand over the building in a phased way permitting both construction and commissioning work to proceed simultaneously on the project. In a planned and managed way the builder will hand over each section of the building to the client who can then begin his or her own work.

The library wanted to examine various ranges of handover dates as there were a number of possible sequences. The team decided to use quantitative risk analysis techniques to predict the probabilities of handover dates for each part of the building. On the basis of risk-based predictions, the final decisions on handover sequence would be made.

Enter James Macrae of Eurolog. This is what was done to examine the phased handovers using quantitative risk management techniques. The risk analysis started off with the construction project plan as prepared by the managing contractor – Laing Management Contractors – as this was downloaded into the Eurolog system. Then the Library's commissioning plan was downloaded from their OpenPlan system and the two simplified into a master plan. To this network plan, risk analysis techniques allow James to do two major things.

Variable durations

Each duration can be replaced with a range of durations associated with likelihoods. You can have either a continuous range or a specific number of durations. James explains a case where a range of durations might be used; Take an inspection task on an area of the building. The optimistic duration might assume that the inspection found nothing outstanding: duration one day. The pessimistic duration might assume that the inspection uncovered a great deal of poor quality work which requires correction: duration six days.

The most likely is that the inspection will reveal a few problems which need tidying up: duration three days. Hence the durations will be one, three or six. Each duration can have a likelihood associated with it. Let's say a 10% likelihood of a six day duration, a 70% likelihood of the three day duration and a 20% likelihood of the one day time frame.

In other types of task – plastering a wall for example – the task might take anything from two to six days depending on the number of plasterers, how well the work goes and all the other things out there waiting to make your project take longer than planned. Hence a duration that can be anything between one and six days.

Variable logic

Here the network plan is brought closer to reality by the addition of alternatives – logical paths that might be followed associated with likelihoods. Let's take a commissioning task like firing up and adjusting the controls on a boiler. The risk team might judge that there is a 30% chance of the commissioning being successful straight off. A second possibility, judged at 40% likely, follows a path where the commissioning is complicated by some problems with the boiler but where simple remedies are available. The final route, only 30% likely, is that the commissioning fails very badly causing the need for extreme remedies.

Each part follows a series of activities which take some time.

Multiple analysis

Given a project in which either or both of these two aspects of variability exist and have been modelled wherever appropriate, you can see that a simple calculation of early and late start and finish dates will not be enough. The system can take the optimistic durations, pessimistic durations, likely and unlikely paths and calculate a number of overall project durations. The system uses a Monte Carlo simulation technique.

The technique relies on random numbers and the name refers to the random selection of numbers in the gaming houses in the principality. The roulette wheel spins, the little ball bearing bounces and a number gets chosen at random.

What happens to the library project plan is that it is analysed many times using durations selected scientifically at random from the range of durations that form part of the model. The risk analysis system takes each task and uses a random number generation technique called the Monte Carlo technique to select a single duration for each task from the available range and to select a single path where choices are available. The system then executes a normal critical path analysis and arrives at an overall project duration. In each specific pass some tasks are assumed to go really well, some really badly and some in between. The program notes down the overall duration and goes back to the start.

Taking another set of randomly generated numbers, the system chooses some more options and generates another overall project duration. Do this a few hundred times and a graph emerges linking overall project duration to the likelihood of that duration occurring.

Instead of one overall duration based on loads of assumptions, we have a large range of durations each based on assumptions and we can deduce some trends.

In the case of the new library rather than a single overall project duration, this use of risk analysis leads to a prediction of possibilities for each of the numerous handover dates – James Macrae's Window Report shows the sensible range for each handover and the likelihood of each occurrence. It looks much like a barchart but it shows a range of dates for each event – the window in time during which the event is likely to happen.

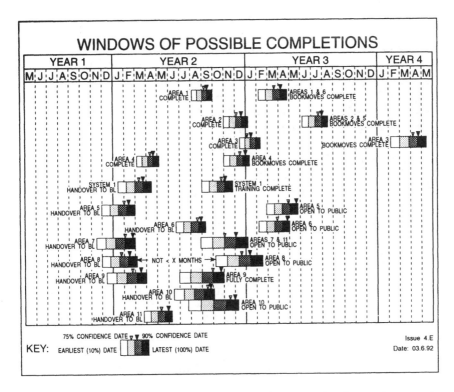

Not only do you get a range of possible durations for the project, each associated with a likelihood of occurring, you also get a valuable tool called a **criticality diagram**. The criticality diagram shows the likelihood of each task being critical. While a normal single pass critical path analysis shows that either a task is or is not critical, the criticality diagram shows a percentage against each task varying from 0% (not critical in any pass) to 100% (critical in every pass) and anything in between.

The technique involved falls under the general heading of **quantitative risk analysis methods**. We are not here concerned with the **qualitative risk analysis techniques** except to say that they exist and perform a useful function in **project risk analysis methods** (PRAM) especially in identifying areas where quantitative techniques are most likely to be useful.

This all gives the client a better understanding about the plan for the building. You can see that once this has all been done, variations and information describing actual work done can be added to the network plan to see their effect on the end date/likelihood graph and the criticality diagram.

There are disadvantages to these techniques. It takes time and resources to work out the numbers so there has to be a significant investment throughout the project management team to facilitate the risk analysis work. There must be a critical path model for the program to work successfully and the various parties involved must contribute to the plan so a degree of openness must exist between the organizations.

The benefits of this approach do not arrive without some considerable commitment and investment. Clearly a project network plan and good estimating are the foundations on which analytical skills to model uncertainties can be used to analyse risk. The information that springs from risk analysis will be useless without decision makers ready to acknowledge and deal with uncertainty.

Sharing the risk

The library involves many organizations contracted to each other but let's think about these techniques being used within a single organization where one group of people are moving towards the start of a new project within the company. In such a case the biggest advantage for me, an old pragmatist, is a personal one which I call risk sharing.

If there are risks associated with a project and you keep them to yourself and tell no one, you take the risk on your own shoulders. If things go well people take little notice and you get off lightly. If things go badly it will be all your fault. If you examine risk, discuss risk with your management and collectively agree to take on the risks then you have done a great job as a project manager. If things go well you get off lightly – just as before. If things go appallingly badly and everything that can go wrong does go wrong, you can say that the organization took sensible, calculated risks. And at least you get to keep your job because you shared the risks.

Some figures from the new British Library

The new British Library, completed in 1995, seats about 1300 readers. If the 11 million books were taken from their shelves and the shelves unwound from their basement walls and laid in a straight line starting at St Pancras heading north, there would be a huge traffic queue in Hendon as the rush hour cars worked their way around this long line of shelving. There would also be a traffic jam in Doncaster and delays in Newcastle where the shelves would end!

The books are stored at a temperature of 17°C and a relative humidity of 50% with very narrow tolerances. These cool, dry conditions apply within some of the air-conditioned exhibition cases.

The Mechanical Book Handling System specifies that 85% of books requested will be delivered within 15 minutes and 95% within 20 minutes.

There is a section of the main library which is a building within a building called the King's Library. It is designed to be a glass 'tower of knowledge' as a visible demonstration of the wealth of information stored in the basements beneath. The King's Library will store a collection of books and manuscripts created by King George III, and given to the nation by King George IV in 1823. This section alone will store 60000 volumes including four folios of Shakespeare, a first edition of *Paradise Lost* and other very early printed works. These are not purely for show as many can be requested and examined.

Contingency and insurance

All this risk analysis stuff does lead somewhere very practical. Once armed with your risk analysis data you can think seriously about contingency and insurance.

A contingency is an amount of money or a period of time set aside in case it is needed. It is sort of part of the predicted costs of the project but it does not relate to the cost of the hinges, headlamps and hired equipment. It is the project management equivalent of a fire extinguisher. You hope that it will sit quietly in its glass box and never be called on to help in a fire. Or perhaps it is like a lifeboat. You take it along in case you need it, but hope you never will. In project management you nearly always do need it desperately.

When something goes over budget or takes too long you can dip into the contingency and liberate a little money. This is discussed a little more fully in part 3 of *Programme Management Demystified* (Reiss, 1996) under the unlikely heading of 'Budgets, cost control and things financial'.

You can see that risk analysis might actually help you to predict better a suitable contingency. You might think that the bigger the contingency the better and in some situations this is true. It is true if you are the project manager and someone in the company has given you the job to get on with. However if you are bidding for a project in a competitive tender you cannot be so generous with the contingency as it has the effect of upping your price thereby making you less competitive.

Also if you are the programme manager who is handing out the job to one of your project managers you might want to find an appropriate (not too tight, not too generous) contingency.

Risk techniques help by giving you some ideas about the things that could go wrong and their likelihoods. When you come to some big risks you can think about getting someone else to take that risk for you. These tend to be the sort of risks that are unlikely but disastrous and this is the business that insurance companies are in. They are risk takers – they take the risk away from you.

This is how it works. You do your risk analysis and decide that fire is a risk on your project. You decide to 'lay off' this risk and take out an insurance policy to protect you. If the project burns down, you get the money back so while you are behind schedule, you might be on budget.

You pay the insurance company a premium for this service and they have experts who calculate the risks and decide how much you should pay them for taking the risk off you. Reading the fine print tells you which risks they are taking and which they are not and gives you a headache from squinting at extremely long words in very small type.

For some reason, insurance companies have a bee in their bonnet about articles falling from aeroplanes and often don't cover these. I am told that from time to time Jumbo jets discharge the contents of the in-flight toilets. I was told that they do this at 9000 metres (30 000 feet) so that the material discharged freezes solid in the cold, high-altitude air. If this is what the insurance companies have in mind it must be because they would themselves die laughing if anyone claimed that their house had been destroyed by 3 tonnes of frozen excrement.

Insurance people lean heavily on a strange and mystical character known as an actuary who devotes his time to working out the probabilities of things going wrong. These are the people who calculate your car insurance premiums because they know the odds of a red MG Metro, driven by a 20-year-old project planner with one conviction and living in Basingstoke, being stolen if it is parked in the street. They do not like or dislike certain cars and people, they deal in statistics and certain kinds of cars and people feature very badly.

Some people, when they talk about risk analysis, mean this world of actuaries, tables, MG Metros and probabilities but this is not the meaning we project management people have in mind at all.

I have been talking here about project risk analysis which is the only kind of risk I know anything about. There are other meanings of risk like the risk analysis exercise carried out by a company thinking of launching a new product to decide if it will be profitable. You'll have to find another book to read about that.

Issue management

We cannot talk about risks without mentioning issues.

A risk is something that *might* impact on the project; an issue is something that *is* impacting on the project.

Issues cause problems with progress, costs and quality. A poor specification may make software design difficult, slow and expensive. Resource scarcity may mean that no one is available to progress some work.

Issues are very often urgent as they hold up progress in one form or another.

Therefore project managers maintain an issue log or an issue register. Each significant problem that cannot be resolved with a phone call should be added to the issue log. Each issue will have an identity, a description, one or more classifications, notes about its impact and measures of its seriousness.

Classifications might include design, prototype, documentation and test. Measures of seriousness might include concern, show-stopper, quality problem and major.

Every issue should have an owner – the name of an individual who is responsible for resolving the issue, for sorting it out. This may be different from the person who raised the issue. Issues may be escalated up the management tree in an effort to find someone with enough authority to overcome the problem.

Each issue should have a status indicator. These often include terms like 'open', 'under review', 'assigned' and 'closed'. Each status indicator may have a date showing when the issue was raised, assigned to an owner and closed.

Terminology

In this chapter are definitions of most of the major terms used by project managers. It is provided as a point of reference in order that potential project managers can refer to and understand the terminology used. The last few pages view some of these popular terms in a jocular but relevant manner.

The first section is dead straight, hopefully rather informative, and the sort of thing that you might occasionally refer to. There is a shortage of definitive definitions in project management. There is a British Standard which you might like to consider, but it does fall short of being the source of all knowledge. Therefore what I have tried to do here is provide workable definitions that will suit most people in most situations. When you hear some term and can't quite recall what it means, you can turn to these pages for guidance.

The second section is also very informative, possibly more than the first half, and it may bring a smile to your face. You've just got to be irreverent sometimes.

Definitions

Activity: An element that takes time and which makes, with other activities, a plan. An activity represents some work being carried out on the project. It is also known as a task.

Activity description: A short definition of the work represented by the activity.

Activity ID: See **Activity name**.

Activity name: A short identifier used to refer to a task.

Activity number: An activity ID restricted to a numeric value.

Activity on arrow plan: A method of representing a project by breaking it down into activities or tasks which are represented by lines meeting at nodes which are represented by circles.

Actual cost: A user-defined indication of how much it has cost to get this far with the particular activity.

Actual finish date: The date that the activity actually finished. This is recorded historically for comparison purposes.

Actual start date: The date that the activity actually started. This is recorded historically for comparison purposes.

Base date: See **Time now date**.

Baseline: See **Target plan** or **Time now date**. Baseline can mean either.

Budget cost: The original estimate of the cost of an activity calculated from the cost of the activity per day and the original duration plus any fixed costs.

Constrained dates: See **Scheduled start date** and **finish date scheduled**.

Cost to complete: The expected cost to complete an activity calculated from the cost of the activity per day and the remaining duration.

Critical activity: An activity which must be completed on time to avoid delaying the end of the project. Critical activities have a free float and a total float of zero.

Critical path: A route through the plan from start to finish through all the critical activities. The path through the plan includes all the activities that you should concentrate on. Delays to the critical path delay the plan end date. There can often be more than one critical path.

Critical path analysis: See **Network analysis**.

Dangle: An activity that is not connected to the plan at both ends. Either the start or the finish or both has no connection to the rest of the plan. It probably indicates a missed relationship or activity – some dangles may be deliberate and some may be in error.

Dependency: see **Link**.

Down time: A period of time when a resource is not available to perform work, or when a project cannot proceed because of holidays or a similar break in production.

Dummy: A logical connection between two activity on arrow nodes showing that one must follow the other. Its name comes from the term **dummy activity** because it differs from an activity only in the fact that no work is going on.

Early finish date (EFD): The earliest possible date the activity may be complete.

Early start date (ESD): The earliest possible date that the activity may begin.

Finish-to-finish link (relationships or **dependencies):** Relationships that connect the finish of one activity to the finish of another. They indicate that an activity cannot end until a specified period of time has elapsed after the end of another activity.

Finish-to-start links (relationships or **dependencies):** Relationships that connect the end of one precedence activity to the start of another. They indicate that one activity cannot start until another has been completed. They are sometimes called simple relationships. (See figure on p. 52.)

Free float: The amount of time that the activity may be delayed without affecting any other activity in the plan. Free float cannot exceed total float.

Histogram: A diagram showing the demand or availability of a resource against a time scale. (See figure on p. 84.)

Key dates: Confusingly key dates have two meanings. One is a milestone (a momentous moment in the project's life). The other meaning is the same as a constraint or scheduled date (something must happen on a specific date). Don't shoot me, I'm just the messenger.

Lag times: The duration of a finish-to-finish relationship which shows by how much the end of one task must leave behind the end of another.

Latest finish date (LFD): The latest possible date by which the activity must be completed if it is not to delay the overall project.

Latest start date (LSD): The latest possible date by which the activity must begin if it is not to delay the overall project.

Lead times: The duration of a start-to-start relationship showing by how much one task should lead another.

Links (also known as *relationships* and **dependencies**): Relationships connect activities in the precedence planning method. Relationships connect precedence activities together to show how they depend on each other. There are three types of relationships.

Loop: A path through a plan that curls around into itself. Such loops are not permitted by network planners.

Milestones: Especially significant points in the project. Activity milestones are events and therefore they take no time and have no duration. They are especially useful in providing summary or overview reports. They are built into the network diagram so that their timing may be calculated.

Monte Carlo analysis/technique: A system for choosing random numbers.

Must dates: See **Scheduled start date** and **finish date scheduled**.

Negative floats: The amount of time you are short of to achieve your objectives for a task.

Network analysis or **network planning:** A method of representing a project by breaking it down into activities. This method is also known as **critical path analysis** and **network analysis** or **PERT**.

Nodes: In activity on arrow plans, nodes represent moments in time. Nodes are used in activity on arrow plans at the start and end of each activity to show the ends of the activities and to show how the activities depend on each other. A start node represents the moment when any activities that come from that node can begin. An end node represents the moment when all activities that lead into that node are complete.

Non-work calendar: This defines special non-working days, e.g. Easter, May Day.

Not after date: See **Scheduled finish date**.

Not before date: See **Scheduled start date**.

Organizational breakdown structure (OBS): A structured method of coding activities into groups in order that they follow the organization of the project.

Original duration: The amount of time that was originally estimated for the activity – often the estimate of duration at the time the project was approved.

Outline, outliner: A tool which allows tasks to be grouped under headings and subheadings. The detail of a part of plan can be expanded and shown or summarized in a single bar.

Percentage complete: The percentage of the work in the activity that has been executed so far. This is used for monitoring and value analysis.

Plan calendar: A working environment that shows days and parts of days that are working and non-working and that relates to all activities in the plan.

Precedence plan: A method of representing a project by breaking it down into activities represented by boxes connected by relationships or by links which are represented by lines.

Priority: A rating given to a task showing how important its needs for resources are in comparison with other tasks. High-priority tasks get resources first through resource smoothing and levelling.

Product breakdown structure: Very much like a **Work breakdown structure**, a PBS shows in a structured hierarchical diagram the components that go into the deliverables of the project.

Progress date: See **Time now date**.

Project finish date: The date on which it is planned to complete the project.

Project milestones: Especially significant points in the project. Milestones are events and therefore they take no time and have no duration. They are especially useful in providing summary or overview reports. Project milestones do not affect the project and are not linked into the network diagram. They serve as a reminder of some special event.

Project start date: The date on which it is planned to start the project.

Qualitative risk analysis: Techniques for listing and weighting risks within a project.

Quantitative risk analysis: A mathematical modelling technique combining critical path analysis and probabilities.

Remaining duration: The amount of time estimated to complete the work that remains to be done in the activity.

Resource: An element that is needed to produce work. It may produce work or may be used in the work. Money is a resource. Any component can be regarded as a resource – some examples of common resources are as follows:

- engineer
- engineer-hour
- labourer
- dumper
- tons of asphalt
- miles of cable
- pavior
- bricks
- concrete.

Resources are sometimes classified by the four Ms:

1. men
2. machines
3. money
4. materials.

Resource hierarchy: A method of indicating how the use of one resource implies the use of other resources. A resource such as a welding gang might imply the use of two welders, one labourer, and one welding rig. The welders, labourers, and welding rigs may also use the resource money.

Resource levelling: A method of adjusting the timing of the activities within a plan to avoid overdemands for resources. An overdemand occurs when demand for a resource exceeds its availability at any time.

Resource profile: A specification showing the availability of a specific resource over the duration of the plan.

Resource scenario: Same as **Resource profile**.

Resource smoothing (optimization): A method of adjusting the timing of the activities within a plan to economize on the use of resources by levelling the demand for each resource.

Risk analysis: Techniques for examining and coping with risks within the project.

Scheduled finish (end) date: A date before which an activity must be completed as a result of some influence outside the project. Scheduled end dates may override the normal timing of the plan. They are used to fix activities in time. Sometimes scheduled is used to mean the date on which the task has been scheduled to occur by the analysis.

Scheduled start date: A date before which an activity cannot begin as a result of some influence outside the project. Scheduled start dates may override the normal timing of the plan. They are used to fix activities in time. Sometimes scheduled is used to mean the date on which the task has been scheduled to occur by the analysis.

Slippage: The excess time required to complete the plan over and above that planned.

Slippage report: A report showing slippage. This report shows two sets of dates for each activity – one being the original timing for the activity and the other being the current timing of the same activity.

Start to start link (relationships or **dependencies):** Relationships that connect the start of one activity to the start of another. They indicate that an activity cannot begin until a specified period of time has elapsed after the start of another activity.

Target plan: A record of the plan at some specific stage used for comparison. The target plan is often laid down when the project is approved or when the tender has been accepted. Later actual work done and plans for future work may be compared with the target.

Time now date: The date from which, for the purposes of the calculation of the plan, the rest of the project starts from. Often tomorrow morning.

Total float: The amount of time that an activity may be delayed without affecting the end date of the project.

Work breakdown structure (WBS): A structured method of coding activities into groups.

Work week calendar: This defines the typical working week.

Alternative definitions

One way to master the terms used in project management is to read some alternative definitions. Much of the truth about project management lies in the truisms that are found in the industry. Here are some definitions that float about the world of project management. Most of these float about in the local bars. I hope they make you smile. I hope even more that they will make you think.

Activity: An element in a plan that involves doing something and which takes time. Usually defined in very short and unclear terms, such as 'Comp n/e grid 34–67 eng work', confusing everyone. If the plan proves to be hopelessly inaccurate or when the inevitable delays occur, it can be redefined at a moment's notice simply by a retranslation of the description.

Free float: The amount of time available for the project manager to sit with his feet on his desk before arriving at the moment of panic. (See **Start node**.)

Start node: The moment when, in theory, the activity can begin. Actually the moment when panic sets in as the project manager realizes that all the necessary machines, personnel, and materials he or she should have accumulated for the task have not yet arrived. Also the moment when excuse preparation begins.

End node: The end of an activity and the beginning of the following activity – usually the allocation of blame. Also the last date for preparation of excuses.

Cost envelope:

(a) The shape contained by the early start cash flow (assuming all activities start as early as possible) and the late start (assuming all activities start as late as possible) cash flow curves.

(b) Any package containing bribes.

Critical task:

(a) Any task that has zero float and that must be done on time to avoid delays to the overall project.

(b) Any task that can be seen from the project director's office window.

Barchart:

(a) A chart showing the activities in the project drawn to a time scale – it looks rather like an uneven staircase drawn from the side view.

(b) A map of local pubs.

Duration: The estimated amount of time that the activity will take – usually between 30 and 50% of the actual time it will take. Duration can be calculated by either a detailed analysis of previous activities of a similar nature and reference to the company database on time-analysis studies, or, more often, by wetting the end of one finger and waving it around in a gentle breeze.

Remaining duration: The amount of time left to spend on a task until its completion. This starts off equal to the original duration of the activity, and, as a result of the optimism shown by project managers, rapidly decreases. Thus begins a lengthy period when the remaining duration is close to, and even approaches, but never actually reaches zero. This is referred to as the persistent 99% complete syndrome and results in the saying that '99% of tasks in 99% of projects are 99% complete for 99% of the time'.

Milestone: A specially selected activity chosen as a key point in the project and used frequently for reporting to senior management.

Traditionally, milestones are carefully selected, given complex sounding descriptions, and positioned way off the critical path. By this means, senior managers can be fooled into believing the project is proceeding on time until near the very end. By this time you will have been promoted to a more senior position on another project and can blame those that took over from you for the disastrously late completion of the project.

PERT: This acronym abbreviates Project Evaluation Review Technique. Originally it was a clever idea to use a pessimistic, an optimistic, and a most likely duration for each activity. In this technique the plan was to be analysed many times, and the result was a Gaussian distribution curve showing overall project duration against likelihood of success. As this involved a lot of work, PERT has come to mean combining the three durations into one by the formula

$$\frac{\text{most likely duration} \times 4 + \text{pessimistic} + \text{optimistic}}{6}$$

Then the plan is analysed during the resulting duration. Hence this is the same as waving three damp fingers in the breeze. (See **Duration**.)

Dummy:
(a) An artificial activity used as a logical link but not indicating any particular element of work proceeding.
(b) Anyone foolish enough to be around long enough to take the blame for the failure of the project.

Ladder network:
(a) A technique of drawing a network to represent a repetitive sequence of work, e.g. the fitting out of the floors of a multistorey block of offices or the building of 100 gearboxes.
(b) A drawing of a path around a building site showing the positions of all the ladders and other routes that can be used to surprise otherwise peacefully smoking artisans.

Resources: Items that produce work (carpenters, excavators, cranes) or that get used in the work – e.g. bricks, pipes, rivets. It is absolutely essential that the plan for the project should cause demand to exceed supply of all significant resources. By this means, delays can always be attributed to resource scarcity. The resource that is always in highest demand at the critical moment is people to take the blame.

Resource histogram: A picture of the Manhattan skyline drawn on a piece of graph paper to substantiate the argument that resource scarcity is causing all the delays.

Critical path:

(a) The longest path through a network; it therefore includes those activities that if delayed will delay the whole project.

(b) The route through the site, factory, or office that is used to show senior management the project; it avoids the late parts of the project and includes all that are on or ahead of plan.

Cash flow curves:

(a) A graphical representation of the flow of cash on the project drawn against time.

(b) The parts of the female anatomy that encourage men to lavish entertainment and expensive gifts.

Project budget:

(a) The amount of money set aside for the project, often broken down into detail by elements of the project.

(b) A figure which the project manager intends to increase by a minimum of 30% with the aid of a carefully structured, well-planned battle with whoever controls the money.

Burst event:

(a) An event on a critical path diagram out of which many links emerge – therefore an event that controls many others.

(b) The day the director explodes because you failed to achieve the expected profit margin on your project. (See figure on p. 203.)

Burst node:

(a) A node on a critical path diagram out of which many links emerge – therefore a node that controls many others.

(b) What you get if you walk into a brick wall.

Contingencies:

(a) Money set aside for the unforeseen.

(b) Those damn things that you should have thought about before so that you are armed with a good excuse, e.g. 'I'm sorry I'm late, sir/dear/mother . . .'

Events:

(a) The start or end point of a task. A task might be *paint bridge*; an event would be *Bridge painting finished*.

(b) Henley, Ascot and Wimbledon.

Logical delays:

(a) Links between tasks showing how one task cannot begin until a certain time after another.

(b) Any believable excuse such as 'The train broke down', 'I sent it by first-class post', and 'I couldn't avoid the M25'.

Activity time:

(a) The estimated duration of a task or activity.

(b) Usually between the hours of 12 and 1 or 1 and 2 p.m.

Forward pass:

(a) An analysis of a critical path diagram passing forwards through

the plan to discover the earliest start and finish dates of the tasks.

(b) Khyber.

Backward pass:

(a) An analysis of a critical path diagram passing backwards through the plan to discover the latest start and finish dates of the tasks.

(b) Rebyhk.

Free float:

(a) The amount of time a task can be delayed without affecting any other tasks.

(b) An asset that money can usually buy. (See **Resources**.)

Subproject:

(a) A part of a project; a neat group of tasks that can be viewed and dealt with together within the overall project.

(b) The type of project that gives the whole team that sinking feeling.

Report runs:

(a) A procedure to produce a whole collection of reports; often many copies of many reports.

(b) The ten-minute dash to the WC just before presenting a report to a major review meeting – usually blamed on 16 pints of bitter and a vindaloo the night before.

Target setting:

(a) The system of setting goals for the team, such as getting the roof on by 1 July, to encourage motivation and understanding of the requirements.

(b) Putting yourself (or, even better, someone else) in line to take the blame for the project's failure.

Finally, we conclude with the five stages of a project (there is more truth in this list than in most project-management books):
1. initial enthusiasm;
2. onset of reality;
3. panic;
4. blame of the innocent;
5. reward for the uninvolved.

Appendix 1: Additional sources of information on project and programme management

Some additional sources of information on programme management:

- Association for Project Management (APM)
  - www.apm.org.uk
- The Project Management Institute (PMI)
  - www.pmi.org
- ProgM – The Programme Management Special Interest Group
  - www.e-programme.com
- The Project And Programme Office Specific Interest Group (PPSO SIG)
  - http://www.pmi-pmosig.org/
  - http://www.psoforum.com/
- UK Office of Government Commerce
  - www.ogc.gov.uk/index.asp?id=38

Index